Geometrical Optics of Weakly Anisotropic Media

Geometrical Optics of Weakly Anisotropic Media

A.A. Fuki
Geophysical Center, Russian Academy of Sciences, Moscow, Russia

Yu. A. Kravtsov
Space Research Institute, Russian Academy of Sciences, Moscow, Russia

O.N. Naida
Novgorod State University, Novgorod, Russia

Translated from the Russian by S.D. Danilov

CRC Press
Taylor & Francis Group
Boca Raton London New York

CRC Press is an imprint of the
Taylor & Francis Group, an **informa** business

CRC Press
Taylor & Francis Group
6000 Broken Sound Parkway NW, Suite 300
Boca Raton, FL 33487-2742

First issued in paperback 2020

© 1998 by Taylor & Francis Group, LLC
CRC Press is an imprint of Taylor & Francis Group, an Informa business

No claim to original U.S. Government works

ISBN-13: 978-0-367-45579-8 (pbk)
ISBN-13: 978-90-5699-036-7 (hbk)

Visit the Taylor & Francis Web site at
http://www.taylorandfrancis.com

and the CRC Press Web site at
http://www.crcpress.com

British Library Cataloguing in Publication Data

Fuki, A.A.
    Geometrical optics of weakly anisotropic media
    1. Anisotropy 2. Geometrical optics
    I. Title II. Kravtsov, IU. A. (IUrii Aleksandrovich)
    III. Naida, O.N.
    535.3'2

In memory of our teacher, the outstanding Russian physicist Professor S. M. Rytov (1908–1996), who was renowned for his work on the geometrical optics method, the theory of oscillations and statistical radiophysics.

Contents

Preface

Interest in weakly anisotropic media was first raised by the problem of the limiting polarization of radio waves propagating in an inhomogeneous plasma. It turned out that the transformation of radio wave polarization in a plasma (ionospheric, solar, laboratory) occurs largely in a relatively thin weakly-anisotropic layer in between the outer isotropic medium (neutral gas or vacuum) and the essentially anisotropic magnetized plasma.

From a physical viewpoint, a noticeable change of polarization in a weakly anisotropic medium is associated with the lifting of polarization degeneration of a transverse electromagnetic wave passing from an isotropic medium to an anisotropic one. Budden (1952) was the first who perceived the physical principles underlying the phenomenon of limiting the polarization of radio waves in the ionosphere. The effective theory of polarization transformation for plane-layered media is due to him (Budden, 1952, 1961).

Our interest in weakly anisotropic media was also stimulated by the problem of limiting the polarization of radio waves in the ionosphere. After the quasi-isotropic approximation of geometrical optics had been put forward (Kravtsov, 1968), our efforts were concentrated on applying this new approach to the solution of specific ionospheric propagation problems that were of practical interest throughout the world at that time, back in the seventies and eighties. The quasi-isotropic approximation (QIA) proved to be more convenient and—most importantly—more versatile as compared to the Budden method. First, the QIA applies to 3D inhomogeneous media in contrast to plane-layered media as in the case of the Budden method. Second, along with bending, the QIA allows for the torsion of rays, as well as bending of external magnetic field lines, bending and twisting of optical and elastic axes, etc.

With time it became clear that the quasi-isotropic approximation could be applied to a wide range of phenomena in weakly anisotropic media, involving change of the polarization of arbitrary vector fields. These include, together with electromagnetic waves in a plasma, light waves in deformed crystals and optic waveguides, acoustic waves in weakly anisotropic elastic media, and even spinor wave functions of particles in inhomogeneous magnetic fields. Diverse and ever expanding applications of the quasi-isotropic approximation encouraged us to write this book.

The authors are grateful to have had the opportunity to work with Professor S. M. Rytov, one of the most brilliant Russian physicists of the twentieth century. Being one of the founders of modern geometrical optics, S. M. Rytov was keenly interested in the quasi-isotropic approximation of geometrical optics. Together with M. L. Leontovich he encouraged us to publish our first works on the QIA without delay, gave a great deal of specific advice on the mathematical presentation of the results and provided unparalleled support for the development and refinement of the QIA.

Our long contact with S. M. Rytov in many respects determined the paths our scientific careers would take and remains a bright memory.

Until very recently, geometrical optics dealt with either transverse electromagnetic waves in isotropic media (Rytov's method) or independent (noninteracting) normal waves in anisotropic media (the Courant–Lax method). Unfortunately, neither method is valid in weakly birefringent media which *de facto* remained outside the main body of geometrical optics (Kravtsov and Orlov, 1990). Meanwhile it is precisely in the regions of weak birefringence that the principal phenomenon—the strong interaction of normal waves—occurs. It leads to a noticeable redistribution of energy between the normal waves and causes a substantial change in the polarization structure of the vector field. The quasi-isotropic approximation is an ideal tool for investigating the polarization transformation of a field in a 3D inhomogeneous media. To date this method has only been described in journal articles and, very briefly, in a monograph by Kravtsov and Orlov (1990). This book is entirely dedicated to the presentation of the QIA—an effective method that bridges the gap between the geometrical optics of isotropic media (the Rytov method) and that of anisotropic media (the Courant–Lax approach).

Looking at the progress made, it is a pleasure to realize that the quasi-isotropic approximation presents the final element of the geometrical optics of anisotropic media. The QIA offers a systematic method for describing waves in weakly anisotropic media, which were until recently completely beyond the scope of geometrical optics, and thus provides a continuous transition from the geometrical optics of polarization degenerate transverse waves in an isotropic medium to the geometrical optics of independent normal waves in an anisotropic medium. Therefore, with respect to a 3D inhomogeneous medium, the QIA serves to solve the problems which Budden had solved in his time with respect to a plane-layered medium. Specifically, this allows us to speak of completing the construction of the architecture of geometrical optics. Its foundations were established back in the eighteenth and nineteenth centuries (Fermat, Huygens, Hamilton, Fresnel, to mention but a few); however, its main levels appeared in this century due to the efforts of Sommerfeld and Runge, Debye, Rytov, Courant and Lax. The quasi-isotropic approximation to be stated below not only combines the 'isotropic' and 'anisotropic' parts of the architecture of geometrical optics by conveniently linking the levels but also exercises an independent role of *the geometrical optics of weakly anisotropic media*. This branch of geometrical optics derives from its own original ideas and has its own specific applications.

The authors believe that a unified approach to the description of waves in isotropic, weakly anisotropic and strongly anisotropic media, offered by the quasi-isotropic approximation of geometrical optics, will be of interest in many fields of physics and will have new practical applications.

Chapter 1 was written by Yu. A. Kravtsov and A. A. Fuki; Chapters 3, 6 and 7 belong to O. N. Naida and Chapter 8, to Yu. A. Kravtsov; Chapter 5 was written by O. N. Naida and Yu. A. Kravtsov and Chapters 2 and 4 were written by all the authors together. The book was edited by Yu. A. Kravtsov.

We would like to express our gratitude to S. D. Danilov for the translation into English, to A. N. Pilipetsky for his cooperation in writing Section 5.3, and also to N. Yu. Komarova and O. O. Zvereva for their very valuable help in preparing the manuscript.

References

Budden, K. G. 1952. The theory of the limiting polarization of radio waves reflected from the ionosphere. *Proc. Roy. Soc A*. **215**(2), 215–233.

Budden, K. G. 1961. *Radio Waves in the Ionosphere*. Cambridge: Cambridge University Press.

Kravtsov, Yu. A. 1968. Quasi-isotropic approximation of geometrical optics. *Doklady AN SSSR*, **183**(1), 74–76 (Engl. transl. in *Sov. Physics - Doklady* 1968).

Kravtsov, Yu. A. and Orlov, Yu. I. 1990. *Geometrical Optics of Inhomogeneous Media*. Berlin Heidelberg: Springer-Verlag.

1

Introduction

1.1 Vector field polarization transformation in weakly anisotropic media

The problem of the propagation of electromagnetic, elastic or other waves in anisotropic media makes up a substantial part of wave physics. The propagation of electromagnetic waves of different bands in a weakly magnetized plasma (laboratory, ionospheric, solar, interstellar), the influence of weak anisotropy of magnetic condensed matter, piezo- and ferromagnets on the transmission of electromagnetic waves, light waves in chiral media and in liquid crystals, polarization phenomena in deformed light guides, propagation of acoustic and seismic waves in weakly anisotropic and/or weakly deformed elastic media, splitting of beams of particles with spin in magnetic fields, that is the list of questions, which is by no means complete, connected with the theory of wave phenomena in weakly anisotropic media.

Melrose and McPhedran (1991), renowned for their contribution to electromagnetic wave theory, define a medium as a weakly anisotropic one if the effect of anisotropy can be treated as a perturbation breaking the degeneracy of the transverse states of polarization. The authors of this book use the the same definition.

Needless to say, the weakness of the anisotropy does not imply the polarization transformation of the electromagnetic field being small. Rather, the contrary holds true: it is precisely media with sufficiently weak anisotropy that are capable of controlling the field polarization structure. To mention a high-school example, let us consider crystal plates capable of converting the linearly polarized light into light with arbitrary elliptical polarization. In particular, a quarter wave plate, i. e., a plate of thickness d in which the difference of optical paths $|n_1 - n_2|d$ between two linearly polarized waves equals $\lambda_0/4$,

$$|n_1 - n_2|d = \frac{\lambda_0}{4} = \frac{\pi}{2k_0} = \frac{\pi c}{2\omega},$$

can convert linearly polarized light into circularly polarized light.

Another example of fundamental change of polarization is furnished by the Faraday rotation occurring in natural active media or in media placed in a magnetic field, for example in a magnetized plasma. In either case intrinsic normal waves are polarized circularly. If n_R and n_L denote the refractive indices, respectively, for the right and left circular polarizations, the medium slab of thickness d rotates the electromagnetic strength vector through the angle

$$\psi = -\frac{1}{2}k_0(n_R - n_L)d$$

(Landau and Lifshitz, 1960; Melrose and McPhedran, 1991).

In both examples, even a weak anisotropy,

$$|n_1 - n_2| \ll 1 \quad \text{or} \quad |n_R - n_L| \ll 1,$$

may' be responsible for substantial changes in the polarization of a vector field across a distance on the order of

$$d \approx \frac{\lambda_0}{|n_1 - n_2|} \quad \text{or} \quad d \approx \frac{\lambda_0}{|n_R - n_L|}.$$

In the examples considered above, the radical transformation of the polarization structure of the field only occurs at boundaries bounding the homogeneous medium. It is there that the degeneracy of transverse polarization states breaks.

Contrary to that, in a weakly anisotropic media whose parameters vary continuously, mutual transformation of normal waves happens at each point with the resultant polarization being determined by the magnitude of anisotropy and the rate at which the medium parameters vary.

The main feature of inhomogeneous weakly anisotropic media is their capability of substantially transforming the polarization structure of vector fields. Polarization transformation in weakly anisotropic regions of an inhomogeneous medium owes its existence to a strong interconversion of normal waves: it is in these regions that the polarization degeneracy smoothly arises or lifts. In essence, in a weakly anisotropic medium, the polarization structure of a wave field undergoes a drastic transformation: while in anisotropic medium the electromagnetic field is represented by a superposition of two normal waves, each characterized by a definite, intrinsic polarization, in an isotropic medium transverse magnetic waves are polarization-degenerate and the state of their polarization is not defined.

On entering an anisotropic medium, a transverse electromagnetic wave ceases to be polarizationally degenerate and experiences a substantial change of polarization. The conversion (reconstruction, "perestroyka") of the polarization structure occurs in a weakly anisotropic layer that matches isotropic and anisotropic media, as shown in Figure 1.1). The final result of the polarization conversion ("polarization per-estroyka") is the transformation of the polarization-degenerate transverse wave into a superposition of independent normal waves.

The polarization transformation of waves in an anisotropic medium is a physical problem of general character, which is important both theoretically and practically. This book is dedicated to its analysis.

1.2   Basic methods of wave field description in weakly anisotropic media

1.2.1   THE BUDDEN METHOD

The final polarization state, or, as it often called, the "limiting" polarization of a vector wave field depends on the character of the inhomogeneities encountered in a weakly anisotropic medium. Describing the process of wave transformation in such a medium presents a rather complex mathematical problem which has not been completely solved up to now.

The theory of the linear interaction of normal modes in a plane-layered plasma was elaborated by Budden (1952, 1961) (the Budden method). In a layered medium, Maxwell's equations convert to a system of four coupled ordinary differential equations. The Budden method, devised for smoothly inhomogeneous media, allows one to reduce substantially the order of that system: it becomes second-order and involves a pair of coupled first-order equations. It is these two equations that describe the linear interaction of normal waves in the framework of the Budden method. The method is not constrained by the requirement of weak anisotropy; however the bulk of applications deal with electromagnetic wave propagation in weakly anisotropic media (weakly magnetized plasmas).

The Budden method is widely used in applied investigations and is presented in many text-books (Ginzburg, 1970; Davis, 1969). A comprehensive analysis of results ensuing from it has been carried out by Zheleznyakov, Kocharovskii, and Kocharovskii (1983).

Initially, the Budden method concerned only electromagnetic waves in a plasma. Later, Budden's results were extended by Cheng and Fung (1977a, b), Fung and Cheng (1977) and Oldano (1987) to arbitrary dielectric layered media.

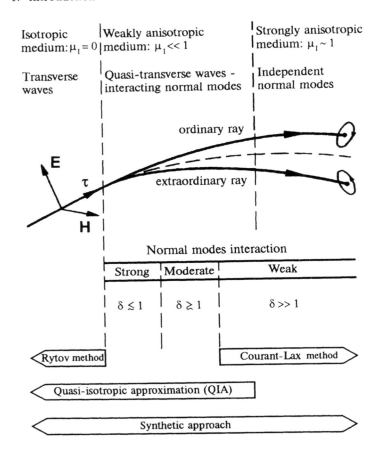

FIGURE 1.1. Transformation of a transverse (polarizationally degenerate) wave into a superposition of independent normal waves in a weakly anisotropic layer between the isotropic and anisotropic media. The wave transformation occurs in the region of strong normal wave interaction, where the parameter δ is less than unity. The parameter $\delta = \mu_1/\mu$ is the ratio of the parameter μ_1 that characterizes the strength of anisotropy to the small parameter μ of geometrical optics. The parameters δ, μ_1 and μ are defined by Eqs. (1.1), (1.3) and (1.4) of Section 1.2. The validity range of the quasi-isotropic approximation (QIA) of geometrical optics covers an isotropic medium and, partially, the region of independent propagation of normal waves in an anisotropic media. In this way the QIA matches the geometrical optics of isotropic media (the Rytov method) with that of anisotropic media (the Courant–Lax method). Dashed line between the rays shows the "isotropic" ray.

1.2.2 QUASI-ISOTROPIC APPROXIMATION (QIA) OF GEOMETRICAL OPTICS

Another approach to the description of waves in inhomogeneous media has been suggested by Kravtsov (1968a). That approach, called the quasi-isotropic approximation (QIA) of geometrical optics, is not constrained by plane-layered media, and applies to arbitrary 3D inhomogeneous media. The general condition of applicability of geometric-optical methods is the requirement that the wavelength λ must be substantially less than the distance l over which either the wavelength itself or the polarization of the wave change noticeably ($\lambda \ll l$). This condition can be conveniently written as the requirement that the geometric-optical parameter is small:

$$\mu \equiv \frac{1}{k_0 l} \ll 1, \tag{1.1}$$

where $k_0 = \omega/c$ is the wave number.

It is not easy to give a simple and simultaneously universal definition to the characteristic scale $l(\mathbf{r})$ of inhomogeneity of the medium that appears in the condition (1.1). Perhaps the only exception is the case of an isotropic medium where the characteristic scale $l_0(\mathbf{r})$ is almost obvious: it is the distance over which the refractive index and, accordingly, the wavelength $\lambda = \lambda_0/n$ vary approximately by a factor of two. Formally,

$$l_0(\mathbf{r}) \sim \frac{n(\mathbf{r})}{|\nabla n(\mathbf{r})|} \sim \frac{|\varepsilon(\mathbf{r})|}{|\nabla \varepsilon(\mathbf{r})|}. \tag{1.2}$$

We shall term the scale $l_0(\mathbf{r})$ the "isotropic" one. As a rule, the wave polarization changes insignificantly on the length $l_0(\mathbf{r})$.

The isotropic scale $l_0(\mathbf{r})$ preserves its meaning in an anisotropic medium, and not only for a weak, but also for a strong anisotropy, as the characteristics of the rate at which the trace $\varepsilon_0(\mathbf{r}) = \frac{1}{3}\mathrm{Tr}\,\hat{\varepsilon}(\mathbf{r})$ of the dielectric permittivity tensor changes. However, in an anisotropic medium there exists another scale $l_p(\mathbf{r})$ characterizing the rate of change of the polarization of a given vector field. The polarization scale $l_p(\mathbf{r})$ should be understood as a distance over which the polarization of a normal wave changes considerably (under a natural additional condition that the direction of normal wave propagation is constant or varies weakly). If $e_i(\mathbf{r})$ denotes one of the variable components of the polarization vector $\mathbf{e}(\mathbf{r})$, the polarization scale $l_p(\mathbf{r})$ can formally be introduced by analogy with Eq. (1.2):

$$l_p(\mathbf{r}) \sim \frac{|e_i(\mathbf{r})|}{|\nabla e_i(\mathbf{r})|}. \tag{1.3}$$

On adopting this definition one does not need to account for polarization changes caused by beating between normal waves: geometrical optics can be developed in terms of normal waves.

To sum up, within the region where normal waves exist, there are two scales characterizing the inhomogeneity of a medium: the isotropic scale $l_0(\mathbf{r})$ and the polarization scale $l_0(\mathbf{r})$. Correspondingly, when using the condition (1.1) one can distinguish between the isotropic small parameter

$$\mu_0 = \frac{1}{k_0 l_0} \ll 1, \tag{1.4}$$

and the polarization small parameter

$$\mu_p = \frac{1}{k_0 l_p} \ll 1. \tag{1.5}$$

It is however more sound to combine these two parameters into one:

$$\mu = \max(\mu_0, \mu_p) = \frac{1}{k_0 l} \ll 1, \tag{1.6}$$

with l implying the least from quantities l_0 and l_p:

$$l = \min(l_0, l_p). \tag{1.7}$$

We shall further adopt this definition for the characteristic scale of the inhomogeneity of the medium.

In the framework of the QIA a medium is assumed to be smoothly inhomogeneous, as in the case of the Budden method, i. e. it is assumed that the geometric-optical parameter is small, or the condition (1.1) is satisfied.

The QIA hinges on the assumption that in the zeroth approximation the electromagnetic field has the same transverse structure as it would have in an isotropic medium. However, in contrast to the case of an isotropic medium, where the polarization of the field vector in a plane perpendicular to the ray is arbitrary, in an anisotropic medium the QIA admits the existence of normal waves whose polarization is uniquely defined by the anisotropy tensor,

$$\nu_{ik} = \varepsilon_{ik} - \varepsilon_0 \delta_{ik}. \tag{1.8}$$

Here ε_{ik} is the tensor of dielectric permittivity, ε_0 is its main isotropic part, for example, $\varepsilon_0 = \frac{1}{3} \operatorname{Tr} \widehat{\varepsilon}$. The weakness of the anisotropy is characterized by the parameter

$$\mu_1 = \max_{i,k} |\nu_{ik}| \ll 1, \tag{1.9}$$

which, in the framework of the QIA, serves as a parameter of asymptotic expansion of the wave field, as does the small geometric-optical parameter μ.

The ratio

$$\delta = \frac{\mu_1}{\mu} \tag{1.10}$$

provides a measure of anisotropy strength. For $\delta \gg 1$, the QIA equations reduce to those for independent normal waves, i.e. to the Courant–Lax equations (Courant and Lax, 1956; Lax, 1957; Courant, 1962), whereas for $\delta \to 0$ they convert to the geometrical optics equations of isotropic medium — the Rytov (1938) method. Thus, the QIA allows one to follow the transition from transverse waves in isotropic media ($\delta \to 0$) to independent normal waves in strongly anisotropic media ($\delta \gg 1$), as demonstrated in Figure 1.1. Quite recently this was regarded as being difficult to attain (Courant, 1962).

Later on the QIA equations were subjected to analysis and generalizations. To solve them, Naida applied perturbation theory methods (Naida, 1970, 1971, 1974). He then suggested the method of splitting ordinary and extraordinary rays, introduced QIA equations for the electric induction vector (Naida 1977a, 1978b), formulated QIA equations for acoustic waves in weakly anisotropic elastic media (1977b, 1978a). Together with Prudkovskii, Naida established the relationship between the QIA equations and the quantum-mechanical equations concerned with splitting of particle beams with different spin states in a magnetic field (Naida and Prudkovskii, 1978).

It turned out that the QIA equations are simpler in form than the Budden equations and are often more convenient in specific calculations, although, by their universality with respect to the degree of anisotropy μ_1, they compare unfavorably with the Budden equations: while the QIA requires two parameters, μ and μ, to be small, the Budden method applies if only the geometric-optical parameter μ is small; the anisotropy parameter, μ_1, can be on the order of unity.

The QIA was helpful in calculating effects of quasi-transverse (with respect to an external magnetic field) electromagnetic wave propagation in 3D inhomogeneous plasma (Kravtsov and Naida, 1976; Naida,

1978b), in analyzing a number of radiowave ionospheric propagation problems (Fuki, 1987a, b, c, 1988, 1990a, b; Kravtsov and Fuki, 1990; Tokar, Rubinstein and Nikitin, 1987; Kharitonova, Fuki and Bukatov, 1989), in elucidating peculiarities of polarization in radiowave scattering in the polar ionosphere (Goryshnik and Kravtsov, 1969; Apresyan and Kravtsov, 1996, 1997). It made it possible to analyze systematically the linear transformation effects of waves which enter 3D inhomogeneous plasma and polarization effects in single-mode light guides.

With the help of the QIA, a number of new phenomena were discovered and described. These include linear wave interaction in the region of a neutral magnetoactive field in a plasma (Zheleznyakov and Zlotnik, 1977), the effect of "tangent conical refraction" (Naida, 1979), the interaction of helical waves in liquid crystals (Zheleznyakov, Kocharovskii and Kocharovskii, 1980). The reader may find a presentation of these effects and other achievements of the QIA in our recent review (Kravtsov, Naida and Fuki, 1996).

1.3 Goals and tasks of the book

The book is aimed at a systematic presentation of different modifications of the QIA as applied to describing various polarization effects in weakly anisotropic media.

We premise a short outline of the general conceptions and main branches of the geometrical optics method (Chapter 2, Sections 2.1 – 2.3) to a rigorous presentation of the QIA (Sections 2.4 and 2.5).

Modifications and generalizations of the QIA are treated in Chapter 3. This chapter contains a conceptually new approach (the method of split rays), which accounts for ray splitting that accompanies the decomposition of the total field into independent normal waves (Section 3.3). The method of split rays has a useful generalization even in strongly anisotropic inhomogeneous media (Section 3.4). As with the Budden method, such a generalization is not restricted to small μ_1, and yet, unlike that method, it enables a description of the wave transformation in arbitrary 3D inhomogeneous media.

Selected chapters are devoted to electromagnetic waves in a magnetoactive plasma (Chapter 4), light waves in deformed media and fibers (Chapter 5), acoustic waves in deformed elastic media (Chapter 6), and to the description of the spinor wave function of particles in a magnetic field (Chapter 7). In all cases the weak anisotropy shows up as a parameter of asymptotic expansion.

It is noteworthy that the idea of considering the anisotropy as a small perturbation had already been formulated by Pauli (1932) in relation to the Dirac equation, in which the anisotropy occurs because of a weak magnetic field. Later on Pauli's ideas were taken further first by Galanin (1942), and then by Rubinow and Keller (1963), although without account for the splitting of polarized particle beams. The spin motion along the unsplit trajectories is given most clearly by Akhiezer and Berestetskii (1965). Unfortunately, these results are still not considered as being common knowledge to experts in electromagnetic wave propagation. Equally, the possibilities of the QIA are far from being exhausted in quantum theory. That is why we found it expedient to dedicate a separate chapter to the behavior of particle spinor wave functions in a magnetic field, treating the problem from the point of view of the QIA.

2

Geometrical optics of inhomogeneous media

2.1 General scheme of the geometrical optics method

2.1.1 THE EIKONAL SUBSTITUTION AND THE EIKONAL EQUATION

We remind the reader of basic ideas of method of the geometrical optics using the example of Maxwell's equations and following, for the most part, books by Landau and Lifshitz (1987), and Kravtsov and Orlov (1990). For simplicity, we restrict ourselves to the case of a monochromatic wave (with time dependence $\exp(-i\omega t)$) in an inhomogeneous gyrotropic stationary medium. The absorption and spatial dispersion will be ignored and the magnetic permeability will be set at unity. Under these conditions, an electric field vector $\mathcal{E}$ obeys the equation

$$k_0^2 \hat{\varepsilon} \mathcal{E} - \text{curl curl } \mathcal{E} = 0 \qquad (k_0 = \omega/c, \quad \varepsilon_{mn}^* = \varepsilon_{mn}), \qquad (2.1)$$

where c is the velocity of light in a vacuum. The main geometric-optical tool is the eikonal substitution,

$$\mathcal{E}(r) = \mathbf{E}(\mathbf{r}) \exp[i\varphi(\mathbf{r})], \qquad (2.2)$$

which enables one to separate fast oscillations of the wave field and relatively slow variations in parameters of medium and associated wave parameters. It is implied that the amplitude $\mathbf{E}(\mathbf{r})$ and the eikonal gradient $\mathbf{k} = \nabla\varphi$, playing the role of wave-vector, vary with $\mathbf{r}$ at a much slower scale than the exponential function $\exp(i\varphi)$.

Constructing the ray asymptotics of a field in a medium requires first of all grouping partial derivatives of basic equations (2.1) into ordinary derivatives along the rays. The eikonal substitution (2.2) reduces Eq. (2.1) to the form:

$$k_0^2 \hat{\varepsilon} \mathbf{E} + \mathbf{k} \times \mathbf{k} \times \mathbf{E} - i(\mathbf{k} \times \text{curl } \mathbf{E} + \text{curl }(\mathbf{k} \times \mathbf{E})) - \text{curl curl } \mathbf{E} = 0, \quad (2.3)$$

where $\mathbf{k} = \nabla\varphi$, $k_0 = \omega/c$.

Impose a quite reasonable condition that the zeroth approximation $\mathbf{E}^{(0)}$ allows a correct transition to the case of a homogeneous medium. Then, in this approximation, one should retain only terms quadratic in $\mathbf{k}$ and k_0 from all terms of Eq. (2.3). That yields a vector equation for $\mathbf{E}^{(0)}$:

$$k_0^2 \hat{\varepsilon} \mathbf{E}^{(0)} + \mathbf{k} \times \mathbf{k} \times \mathbf{E}^{(0)} = 0, \tag{2.4}$$

which is equivalent to the system of three linear equations for components $E_x^{(0)}$, $E_y^{(0)}$, $E_z^{(0)}$. As is well-known, the solvability condition for that system leads to the eikonal equation:

$$\det(k_0^2 \varepsilon_{mn} + k_m k_n - \delta_{mn} \mathbf{k}^2) = 0 \qquad (k_m = \partial\varphi/\partial x_m). \tag{2.5}$$

The ratio

$$n(\mathbf{r}, \mathbf{k}/|\mathbf{k}|) = |\mathbf{k}|/k_0$$

serves as the refractive index and, in turn, satisfies the algebraic equation

$$\det\left(n^{-2}\varepsilon_{mn} + t_m t_n - \delta_{mn}\right) = 0, \tag{2.6}$$

where $\mathbf{t} = \mathbf{k}/|\mathbf{k}|$ is a unit vector aligned with the wave vector. For $t_x = t_y = 0$, $t_z = 1$ the known formula follows:

$$n_{1,2}^{-2} = \frac{1}{2}(\chi_{xx} + \chi_{yy}) \pm \left[\frac{1}{4}(\chi_{xx} - \chi_{yy})^2 + |\chi_{xy}|^2\right]^{1/2}, \tag{2.7}$$

where $\chi_{xx}, \ldots$ stand for the components of the inverse tensor of dielectric permittivity $\hat{\chi} = \hat{\varepsilon}^{-1}$ (Landau and Lifshitz, 1960).

For an isotropic medium Eqs. (2.4)–(2.7) can be substantially simplified:

$$(k_0^2 \varepsilon - \mathbf{k}^2)\mathbf{E}^{(0)} + \mathbf{k}(\mathbf{k} \cdot \mathbf{E}^{(0)}) = 0, \tag{2.4a}$$

$$k_0^2 \varepsilon - \mathbf{k}^2 = 0, \tag{2.5a}$$

$$n^{-2}\varepsilon - 1 = 0, \tag{2.6a}$$

$$n_1 = n_2 = n = \varepsilon^{1/2}. \tag{2.7a}$$

2.1.2 HAMILTON'S EQUATIONS FOR RAYS

Eq.(2.5) is called the dispersion equation. Viewed as an algebraic equation with respect to frequency $\omega = k_0 c$, the local dispersion relation (2.5) has, as a rule, among its roots two positive ones:

$$\omega = \Omega_{1,2}(\mathbf{r}; \mathbf{k}). \tag{2.8}$$

In an anisotropic medium, $\Omega_1 \neq \Omega_2$ except for special directions. We shall assume that $\Omega_1 > \Omega_2$, with Ω_1 and Ω_2 corresponding to the extraordinary and ordinary waves, respectively. In an isotropic medium both roots coincide:

$$\Omega_1(\mathbf{r}; \mathbf{k}) = \Omega_2(\mathbf{r}; \mathbf{k}) = \varepsilon^{-1/2} c \, |\mathbf{k}| \, . \tag{2.9}$$

For both isotropic and anisotropic media, the rays that correspond to the zeroth approximation in the form of (2.4) and (2.5) obey the Hamiltonian system of equations

$$\frac{d\mathbf{r}}{dt} = \frac{\partial \Omega_a}{\partial \mathbf{k}},$$
$$\qquad\qquad (a = 1, 2) \tag{2.10}$$
$$\frac{d\mathbf{k}}{dt} = -\frac{\partial \Omega_a}{\partial \mathbf{r}},$$

where t is the time and $\partial \Omega_a / \partial \mathbf{k}$ is the group velocity. Here a is the wave polarization index, say $a = 1$ for the extraordinary wave and $a = 2$ for the ordinary one.

In a stationary case Eqs.(2.10) can be written in the form:

$$\frac{d\mathbf{r}}{ds} = \frac{\beta_a(\mathbf{r}, \mathbf{k})}{|\beta_a(\mathbf{r}, \mathbf{k})|}, \quad ds = |d\mathbf{r}|, \quad \beta_a = \frac{\partial \, |\mathbf{k}| \, n_a^{-1}(\mathbf{r}, t)}{\partial \mathbf{k}},$$
$$\tag{2.11}$$
$$\frac{d\mathbf{k}}{ds} = -\frac{|\mathbf{k}|}{|\beta_a(\mathbf{r}, \mathbf{k})|} \frac{\partial n_a^{-1}(\mathbf{r}, t)}{\partial \mathbf{r}} \quad (a = 1, 2; \, \mathbf{t} = \mathbf{k}/|\mathbf{k}|).$$

The refractive indices $n_a(\mathbf{r}, t) = |\mathbf{k}_a| / k_0$ are the positive solutions of Eq.(2.6), and $\tau = d\mathbf{r}/ds$ is a unit vector tangent to the ray. Note that the unit vector $\mathbf{t} = \mathbf{k}/|\mathbf{k}|$ may not coincide with the tangent unit vector τ, see Figure 2.1.

In an isotropic medium, Eqs. (2.11) referring to different values of a do coincide. They admit a simplification in accordance with Eq. (2.9):

$$d\mathbf{r}/ds = \mathbf{k}/ \, |\mathbf{k}| \, , \qquad d\mathbf{k}/ds = k_0 \nabla n. \tag{2.12}$$

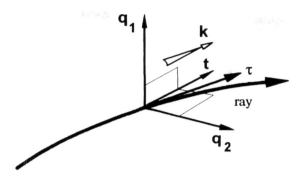

FIGURE 2.1. Right real system of unit vectors q_1, q_2, and t connected with a ray, and the tangent unit vector τ. In an anisotropic medium the vector τ differs from the vector t

The tangent unit vector to the ray τ in this case coincides with the unit vector t.

According to (2.10) and (2.11), the eikonals φ_1 and φ_2 of the extraordinary and ordinary waves are given by formulae

$$\varphi_a = \int \mathbf{k}_a \, d\mathbf{r}_a = \int \mathbf{k}_a (\partial \Omega_a / \partial \mathbf{k})|_{\mathbf{k}=\mathbf{k}_a} \, d\mathbf{r}_a,$$

$$\mathbf{r}_a = \int |\boldsymbol{\beta}_a(\mathbf{r}_a, \mathbf{k}_a / |\mathbf{k}_a|)|^{-1} \boldsymbol{\beta}_a(\mathbf{r}_a, \mathbf{k}_a / |\mathbf{k}_a|) \, ds_a,$$

(2.13)

where ds_a is the element of ray length (2.11), and functions $\mathbf{r}_a(s_a)$ and $\mathbf{k}_a(s_a)$ are also defined by Eqs. (2.11). Eq. (2.13) becomes substantially simpler in the case of an isotropic medium:

$$\varphi = k_0 \int \varepsilon^{1/2} \, ds = k_0 \int n \, ds.$$

(2.14)

Thus all formulae for rays, eikonals, and refractive indices related to ordinary and extraordinary rays in an anisotropic media transform continuously to their respective "isotropic" counterparts at a continuous transition from the anisotropic to isotropic media. Such a continuity is not exhibited by the equations for wave amplitudes taken from generally accepted theories of geometrical optics, i.e., on the one hand, by the

Courant–Lax equations for anisotropic media, and, on the other, by the
Rytov equations. Considerable departures already arise in zeroth-order
formulae as we shall illustrate below, in Section 2.3.

2.1.3 INDEPENDENT NORMAL WAVES

In the case of an anisotropic medium, in the zeroth approximation, Eq.
(2.4) uniquely defines the polarization of each of the solutions $\mathcal{E}_1^{(0)}$ and
$\mathcal{E}_2^{(0)}$ of Eq. (2.1) (or each of the solutions $\mathbf{E}_1^{(0)}$ and $\mathbf{E}_2^{(0)}$ of Eq. (2.3))
that correspond, respectively, to extraordinary or ordinary waves:

$$\mathbf{E}_a^{(0)} = C_a \mathbf{e}_a, \quad \mathcal{E}_a^{(0)} = C_a \mathbf{e}_a \exp(i\varphi_a) \quad (a = 1, 2). \tag{2.15}$$

Polarization vectors $\mathbf{e}_1$ and $\mathbf{e}_2$ entering (2.15) are functions of arguments $\mathbf{r}$, $\mathbf{k}$, in conformity with vector equation (2.4). Polarization vec-
tors can be normalized, say, according to the condition $|\mathbf{e}_a|^2 = 1$ which
bounds the total length of vectors; however it is more convenient here
to norm a transverse (relative to $\mathbf{t}$) component:

$$|\mathbf{e}_{a\perp}|^2 = |\mathbf{e}_a - \mathbf{t}(\mathbf{e}_a \mathbf{t})|^2 = 1. \tag{2.16}$$

For a Hermitian tensor $\hat{\varepsilon}$ ($\varepsilon_{mn} = \varepsilon_{nm}^*$), expressions for $\mathbf{e}_1$ and $\mathbf{e}_2$ are
the simplest in a right orthogonal system with real unit vectors $\mathbf{q}_1$, $\mathbf{q}_2$,
and $\mathbf{t}$ (Figure 2.1), in which

$$\mathrm{Re}\,(\mathbf{q}_1, \hat{\chi}\mathbf{q}_2) = \mathrm{Re}\,(\mathbf{q}_2, \hat{\chi}\mathbf{q}_1) = 0. \tag{2.17}$$

In this case, as is known, the components of polarization vectors $\mathbf{e}_1$ and
$\mathbf{e}_2$ are expressed by the formulae (Landau and Lifshitz, 1960):

$$\mathbf{e}_1 = \hat{\chi}\mathbf{d}_1 = e_{11}\mathbf{q}_1 + e_{12}\mathbf{q}_2 + e_{13}\mathbf{t},$$

$$\mathbf{e}_2 = \hat{\chi}\mathbf{d}_2 = e_{21}\mathbf{q}_1 + e_{22}\mathbf{q}_2 + e_{23}\mathbf{t}, \tag{2.18a}$$

whereas the components of electric induction vector $\mathbf{d}$, corresponding
to polarization vectors $\mathbf{e}_1$ and $\mathbf{e}_2$ are strictly transverse, $\mathbf{dt} = 0$, and
are given by expressions

$$\mathbf{d}_1 = (e_{11}\mathbf{q}_1 + e_{12}\mathbf{q}_2)n_1^2,$$

$$\mathbf{d}_2 = (e_{21}\mathbf{q}_1 + e_{22}\mathbf{q}_2)n_2^2, \tag{2.18b}$$

where refractive indices $n_{1,2}$ are defined by (2.7). The coefficients e_{ij}
stand for the combinations

$$e_{11} = e_{22} = (1 + K_1^2)^{-1/2}, \quad e_{12} = e_{21} = -iK_1(1 + K_1^2)^{-1/2},$$

$$e_{13} = n_1^2(\chi_{31}e_{11} + \chi_{32}e_{12}), \quad e_{23} = n_2^2(\chi_{31}e_{21} + \chi_{32}e_{22}),$$

$$(2.19)$$

where the rest of notation is as follows:

$$\chi_{mn} = (\mathbf{q}_m, \widehat{\chi}\mathbf{q}_n) \quad (m, n = 1, 2, 3),$$

$$K_1 = JK^J, \quad J = -\operatorname{sgn} \operatorname{Im} \chi_{12}, \quad K = Q - (1 + Q^2)^{1/2}, \qquad (2.20)$$

$$Q = \frac{i(\chi_{22} - \chi_{11})}{2\chi_{12}}.$$

In the case of a non-gyrotropic birefringent medium, for which $Q \to \infty$, we shall simply obtain:

$$e_{11} = e_{22} = 1, \quad e_{12} = e_{21} = 0$$

instead of the first two formulae (2.19). Expressions (2.18a) satisfy the orthonormality condition (2.16) for the electric polarization vectors

$$e_{ax}^* e_{bx} + e_{ay}^* e_{by} = \delta_{ab} \quad (a, b = 1, 2), \qquad (2.21)$$

together with the condition

$$\mathbf{e}_a^* \cdot \mathbf{d}_b = n_a^2 \delta_{ab} \quad (a, b = 1, 2), \qquad (2.22)$$

where δ_{ab} is the Kronecker delta.

The classical method presented here is intended for anisotropic media and assumes the existence of two independent normal waves (2.15). Such an approach was rigorously implemented by Courant and Lax (1956), Lax (1957), Courant (1962). We describe it in Section 2.3.

The polarization of each of the normal waves strictly follows the rotations of the medium anisotropy axes given by the tensor $\widehat{\varepsilon}(\mathbf{r})$. As shown already by Budden (1952, 1961) in the example of a plane layer, the normal wave approximation reasonably fits the exact solution provided only that the birefringence is strong,

$$\delta = \mu_1/\mu \approx |n_2 - n_1|k_0 l \gg 1, \qquad (2.23)$$

and is violated in a weakly birefringent medium where

$$\delta = \mu_1/\mu \approx |n_2 - n_1|k_0 l \lesssim 1. \qquad (2.23a)$$

Therefore, the Courant-Lax method is, *in principle*, incapable of describing the polarization of a wave leaving an anisotropic region of a medium to enter an isotropic medium (the problem of limiting polarization). We note that Courant (1962) himself pointed at the existence of this problem. Apparently, in the case of an isotropic medium normal waves lose their meaning even as symbols of waves.

2.1.4   THE POLARIZATION STRUCTURE OF A FIELD IN AN ISOTROPIC MEDIUM

In the case of an isotropic medium, Eq. (2.4) no longer uniquely defines the orientation of the vector amplitude $\mathbf{E}^{(0)}$, which corresponds to the polarization degeneration. From Eq. (2.4) it only follows in this case that

$$\mathbf{E}^{(0)} = C_1\mathbf{q}_1 + C_2\mathbf{q}_2, \qquad \boldsymbol{\mathcal{E}}^{(0)} = (C_1\mathbf{q}_1 + C_2\mathbf{q}_2)\exp\left(ik_0\int n\,ds\right),$$

$$\qquad (2.24)$$

where $\mathbf{q}_1$ and $\mathbf{q}_2$ are arbitrary linearly independent vectors which are perpendicular to the tangent $\boldsymbol{\tau} = \mathbf{t}$ (Figure 2.1). In particular, one may take the normal and binormal vectors to the ray, respectively, as $\mathbf{q}_1$ and $\mathbf{q}_2$ (Rytov, 1938):

$$\mathbf{q}_1 = \mathbf{n}, \quad \mathbf{q}_2 = \mathbf{b}, \qquad (2.25)$$

and then

$$\mathbf{E}^{(0)} = E_n\mathbf{n} + E_b\mathbf{b}, \qquad \boldsymbol{\mathcal{E}}^{(0)} = (E_n\mathbf{n} + E_b\mathbf{b})\exp\left(ik_0\int n\,ds\right). \quad (2.26)$$

Thus, in the case of an isotropic medium, the general expression for the zeroth approximation (2.26) differs drastically from its form (2.15) for independent normal waves. This formal difference between Eqs. (2.15) and (2.26) complicates a smooth conjunction of polarization-degenerate transverse waves in an isotropic medium with independent normal waves in an anisotropic medium so much that it was long believed to be impossible (Courant, 1962).

However, a smooth matching of waves propagating from an isotropic medium into an anisotropic one might be obtained in the framework of the quasi-isotropic approximation (QIA) which, on the one hand, preserves the transverse structure of the field in the isotropic medium, and, on the other, allows for a transition to independent normal waves in an essentially anisotropic medium ($\delta \gg 1$).

In order to correlate different, at first glance even incompatible, variants of the ray theory it is convenient to cast Maxwell's equations in a form from which all known variants of the ray method would readily emerge. Daring to overestimate to some extent the significance of the suggested approach and yet not willing to refuse a relevant term, we call the procedure to be stated below as a *universal* one, especially as it also leads to some new results, which cannot be derived in the frameworks of the existing methods.

2.1.5 UNIVERSAL GEOMETRIC-OPTICAL PROCEDURE.

Despite some difficulty caused by the consideration of a 3D inhomogeneous medium, the construction of ray solutions can be achieved by recognizing the fact that the vector of electric induction $\mathcal{D} = \hat{\varepsilon}\mathcal{E}$ is almost transverse (it is strictly transverse in the zeroth approximation) to the wave-vector $\mathbf{k}$. That transversality takes place both in isotropic and anisotropic media. Therefore, in Eq. (2.3), it is natural to use the right trihedron of unit vectors $\tilde{\mathbf{q}}_1, \tilde{\mathbf{q}}_2$, and $\mathbf{t}$, such that $\mathbf{t} = \mathbf{k}/|\mathbf{k}|$, i.e., $\mathbf{t}$ is as previously parallel to the wave-vector whereas $\tilde{\mathbf{q}}_1$ and $\tilde{\mathbf{q}}_2$ are some orthogonal vectors transverse to $\mathbf{t}$ which do not need to be real:

$$(\tilde{\mathbf{q}}_m^*, \tilde{\mathbf{q}}_n) = \delta_{mn} \quad (m, n = 1, 2). \tag{2.27}$$

In all cases we shall assume that the vectors $\mathrm{Re}\,\tilde{\mathbf{q}}_1, \mathrm{Re}\,\tilde{\mathbf{q}}_2$ form a right trihedron together with $\mathbf{t}$.

Substitute

$$\mathbf{E} = E_1\tilde{\mathbf{q}}_1 + E_2\tilde{\mathbf{q}}_2 + E_3\mathbf{t} \tag{2.28}$$

in Eq. (2.3) and introduce the notation

$$\mathbf{D} = D_1\tilde{\mathbf{q}}_1 + D_2\tilde{\mathbf{q}}_2 + D_3\mathbf{t}. \tag{2.29}$$

Having inserted (2.28) into Maxwell's equations (2.3) we scalarly multiply them by some vectors $\mathbf{Q}_1$ and $\mathbf{Q}_2$ which nave not been defined yet (these even do not need to be perpendicular to the ray) and also by the unit vector $\mathbf{t}$. Then we find

$$\mathbf{Q}_1(\widehat{\varepsilon}\mathbf{E} + n^2\mathbf{t} \times \mathbf{t} \times \mathbf{E}) - ik_0^{-1}\mathbf{Q}_1(n\mathbf{t} \times \operatorname{curl}\mathbf{E} + \operatorname{curl}(n\mathbf{t} \times \mathbf{E}))$$

$$-k_0^{-2}\mathbf{Q}_1 \operatorname{curl}\operatorname{curl}\mathbf{E} = 0,$$

$$(2.30a)$$

$$\mathbf{Q}_2(\widehat{\varepsilon}\mathbf{E} + n^2\mathbf{t} \times \mathbf{t} \times \mathbf{E}) - ik_0^{-1}\mathbf{Q}_2(n\mathbf{t} \times \operatorname{curl}\mathbf{E} + \operatorname{curl}(n\mathbf{t} \times \mathbf{E}))$$

$$-k_0^{-2}\mathbf{Q}_2 \operatorname{curl}\operatorname{curl}\mathbf{E} = 0,$$

$$(2.30b)$$

$$D_3 - ik_0^{-1}\mathbf{t}(n\mathbf{t} \times \operatorname{curl}\mathbf{E} + \operatorname{curl}(n\mathbf{t} \times \mathbf{E})) - k_0^{-2}\mathbf{t}\operatorname{curl}\operatorname{curl}\mathbf{E} = 0. \quad (2.31)$$

Formally, relations (2.30) and (2.31) include terms of the zeroth ($k_0^0 = 1$), first, and second orders in powers of the wave number k_0, but, in actual fact, these relations are composed of three groups of terms that differ by the order of smallness with respect to the dimensionless geometric-optical parameter $\mu = 1/k_0 l$. To be certain of this, we note that k_0^{-1} enters (2.30), (2.31) together with derivatives of $\mathbf{E}$ and the first-order medium parameters whereas k_0^{-2} always comes with second-order derivatives. In order of value, these derivatives are assessed as

$$\left|\frac{\partial E_m}{\partial x_n}\right| \sim \frac{|E_m|}{l}, \qquad \left|\frac{\partial^2 E_m}{\partial x_n \partial x_j}\right| \sim \frac{|E_m|}{l^2},$$

where l is the typical scale of the variability of parameters of the medium. Correspondingly, k_0^{-1} always enters as a factor by a quantity of l^{-1} order, and k_0^{-2} is always multiplied by a quantity scaled as l^{-2}. This provides grounds for expanding the amplitude $\mathbf{E}$ in a formal asymptotic series in inverse powers of k_0,

$$\mathbf{E} = \mathbf{E}_0 + \frac{\mathbf{E}_1}{k_0} + \frac{\mathbf{E}_2}{k_0^2} + \cdots,$$

or, equivalently, in the small parameter μ:

$$\mathbf{E} = \mathbf{E}_0 + \mu\mathbf{E}_1 + \mu^2\mathbf{E}_2 + \cdots.$$

Substituting the asymptotic expansions in (2.30) and (2.31) and equating zero coefficients at powers of k_0^{-1} or $\mu = (k_0 l)^{-1}$ yields the eikonal equation in the zeroth approximation and equations for complex amplitudes in the first approximation. Relevant procedures for isotropic media (the Rytov method) and strongly anisotropic media (the Courant–Lax method) are described, for example, by Kravtsov and Orlov (1990). They are presented briefly below, in Sections 2.2 and 2.3.

The choice of vectors $q_{1,2}$ and $Q_{1,2}$ is dictated by considerations of convenience. For an isotropic medium, this may be appropriately done by taking unit vectors of the normal n and binormal b to the ray for unit vectors q_1, Q_1 and q_2, Q_2: $q_1 = Q_1 = n$, $q_2 = Q_2 = b$. Then Eqs. (2.30) allow one to find two amplitudes E_n and E_b characterizing the polarization state of a transverse (polarization degenerate) electromagnetic wave in an isotropic medium. Such a procedure is used in the Rytov method presented in Section 2.2.

In a strongly anisotropic medium, projecting Maxwell's equations, first, on a polarization vector e_1^* associated with the first normal wave and, then, on a polarization vector e_2^* associated with the second normal wave, i.e., on $Q_1 = e_1^*$ and $Q_2 = e_2^*$, yields two equations for complex amplitudes of normal waves. This procedure underlies the Courant–Lax method (Section 2.3).

The quasi-isotropic approximation (Kravtsov, 1968a) mainly differs from the Rytov and Courant–Lax methods in that the QIA classifies small (in weakly anisotropic media) components of the anisotropy tensor

$$\nu_{mn} = \varepsilon_{mn} - \varepsilon_0 \delta_{mn}$$

as being of the first order.

In the framework of the approach developed here (Section 2.4) the "isotropic" set of unit vectors $q_1 = Q_1 = n$, $q_2 = Q_2 = b$ leads to the standard form of QIA equations (Section 2.4.2).

The smallness of the longitudinal component D_3 explicitly follows from (2.31). This explicitness is the advantage of the relations (2.30) and (2.31) as compared with the original form of Maxwell's equations (2.3). As a result we get a reason to exclude the transverse component $D_3^{(0)}$ from the zero approximation and rewrite (2.28), (2.29) in the form

$$\mathbf{D}^{(0)} = D_1 \tilde{q}_1 + D_2 \tilde{q}_2, \quad \mathbf{E}^{(0)} = \hat{\chi}(D_1 \tilde{q}_1 + D_2 \tilde{q}_2). \tag{2.32}$$

Then Eqs.(2.30) lead to a system of equations with two unknown amplitudes D_1 and D_2. The modification of QIA involving the components of the induction vector offers noticeable advantages over the standard QIA form because it possesses a wider region of applicability (Section 3.1).

To proceed we note that the set of complex unit vectors $q_{1,2} = e_{1,2}$ and $Q_{1,2} = e_{1,2}^*$ corresponding to approximate ("simplified") normal waves brings the QIA equations to the form of a system of coupled equations for amplitudes of normal waves (Section 3.2). In this approach, as in the standard QIA form, splitting of the rays associated with both modes should not be large.

If the rays referring to both normal modes split appreciably in the interaction region, the procedure of choosing the unit vectors $q_{1,2}$ and $Q_{1,2}$ becomes more involved than that described above. Along with the conventional vectors $q_1^e = Q_1^{e*} = e_1(k_1)$ and $q_2^o = Q_2^{o*} = e_2(k_2)$ corresponding to extraordinary and ordinary modes, one has to introduce additional polarization vectors $q_2^e = Q_2^{e*} = e_2(k_1)$ and $q_1^o = Q_1^{o*} = e_1(k_2)$ for emerging polarizations. Choosing $q_1 = Q_1^* = q_1^e = e_1(k_1)$ and $q_2 = Q_2^* = q_2^e = e_2(k_1)$ and projecting Maxwell's equations according to (2.30), one obtains a description of one of the split modes. Similarly the choice $q_1 = Q_1^* = q_1^o = e_1(k_2)$ and $q_2 = Q_2^* = q_2^o = e_2(k_2)$ provides a description of the other split mode. Thus the complete picture of wave splitting is described by the system of four equations which are linked pairwise: one pair for the first mode and the other one for the second mode.

We note that the "crossed" unit vectors $e_1(k_2)$ and $e_2(k_1)$ were first introduced by Courant (1962) when deriving small corrections to the polarization of normal waves. The procedure presented here makes use of the crossed polarization vectors in the leading order of the method. It therefore allows one to tackle arbitrary (not necessarily small as in the case considered by Courant) changes of polarization already in the zeroth order of the geometrical optics method. It results in *the split ray method* in the case of a weak anisotropy (Section 3.3) and in *the synthetic approach* in the case of a strong anisotropy (Section 3.4).

Thus, by projecting (2.30)–(2.31) Maxwell's equations on various systems of unit vectors one can embrace all feasible branches of the geometrical optics method, and this demonstrates the universality of this procedure.

When suggesting the universal procedure we touch on a very important, although little explored, question of the completeness of the system of rays and vectors Q_1 and Q_2 corresponding to them. Needless to say, projecting Maxwell's equations (2.3) on the unit vectors Q_1 and Q_2 referring to one or another family of rays involves a certain arbitrariness. Which of the vector pairs Q_1 and Q_2 and which family of rays offer the "best" solution to Maxwell's equations? What is discarded when passing to a geometric-optical solution?

Unfortunately, in relation to the heuristic methods which are used in geometrical optics, such questions still remain unanswered. The reasoning behind these methods hinges mostly on a common sense rather than on rigorous first principles. In this book, when choosing a system of rays and corresponding unit vectors $\mathbf{Q}_{1,2}$, we will basically resort to heuristic arguments.

Luckily, a heuristic ray approach commonly guarantees, in addition to a clear structure, a reasonably high accuracy of the solution even when values of "small" parameter μ are not small. In essence, the main deficiency of geometrical optics is the neglect of diffraction effects. The notion of diffraction effects, generally speaking, possesses some ambiguity. Taken most broadly, these embody everything that makes the difference between the exact and geometric-optical solutions, with the region of applicability of the latter being bounded by the Fresnel criteria (see Kravtsov and Orlov, 1980, 1981, 1990, 1993). Geometrical optics satisfactorily describes the field along the basic rays for smoothly inhomogeneous media. However it completely ignores exponentially weak scattered fields of the order

$$\exp\left(-k_0 l\right) \approx \exp\left(-1/\mu\right).$$

The specific character of these fields is that in the geometric-optical limit $\mu \to 0$ or $k_0 l \to \infty$ they decay stronger than any finite power of small parameters μ or $1/(k_0 l)$. Such fields are not described by the geometric-optical treatment since the latter only accounts for terms with finite powers in μ.

2.2   Geometrical optics of isotropic media (the Rytov method)

In the case of an isotropic medium ($\mathbf{D} = \varepsilon\mathbf{E}$) the zeroth approximation (2.32) takes the form (2.26), i.e. in this case $\widetilde{\mathbf{q}}_1 = \mathbf{n}$ (normal to the ray) and $\widetilde{\mathbf{q}}_2 = \mathbf{b}$ (binormal to the ray). It is also natural to choose $\mathbf{Q}_1 = \mathbf{n}$ and $\mathbf{Q}_2 = \mathbf{b}$. Rytov (1938) discovered that, under the conditions specified, partial derivatives in Eq. (2.30) (with $E_3 = 0$) can be grouped into full derivatives along the ray (2.12):

$$
\begin{aligned}
&\mathbf{n}(nt \times \operatorname{curl} \mathbf{E}^{(0)} + \operatorname{curl}(nt \times \mathbf{E}^{(0)})) \\
&\equiv -2n\frac{dE_n}{ds} - E_n\frac{dn}{ds} - E_n n \operatorname{div} \mathbf{t} + 2nT^{-1}E_b, \\
\\
&\mathbf{b}(nt \times \operatorname{curl} \mathbf{E}^{(0)} + \operatorname{curl}(nt \times \mathbf{E}^{(0)})) \\
&\equiv -2n\frac{dE_b}{ds} - E_b\frac{dn}{ds} - E_b n \operatorname{div} \mathbf{t} - 2nT^{-1}E_n,
\end{aligned}
\tag{2.33}
$$

where ds is the element of length of a ray satisfying (2.12), and T^{-1} is the torsion of the ray (Korn and Korn, 1977, Section 17.2-3):

$$T^{-1} = \mathbf{b} \frac{d\mathbf{n}}{ds}. \tag{2.34}$$

In deriving identities (2.33) we made use of relationships from vector analysis:

$$\mathbf{n} \operatorname{curl} \mathbf{n} + \mathbf{b} \operatorname{curl} \mathbf{b} - \mathbf{t} \operatorname{curl} \mathbf{t} = -2\mathbf{b} \frac{d\mathbf{n}}{ds},$$

$$\mathbf{b} \operatorname{curl} \mathbf{n} - \mathbf{n} \operatorname{curl} \mathbf{b} = \operatorname{div} \mathbf{t}.$$

As a result, the system of two ordinary differential equations follows for the zero-order amplitudes E_n and E_b (we omit the upper index "0" for brevity) from (2.30) :

$$\frac{dE_n}{ds} - \frac{E_b}{T} + E_n \left(\frac{d \ln \varepsilon^{1/4}}{ds} + \frac{1}{2} \operatorname{div} \mathbf{t} \right) = 0,$$

$$\frac{dE_b}{ds} + \frac{E_n}{T} + E_b \left(\frac{d \ln \varepsilon^{1/4}}{ds} + \frac{1}{2} \operatorname{div} \mathbf{t} \right) = 0. \tag{2.35}$$

The last terms in each of the equations of system (2.35) can easily be eliminated by introducing normalized amplitudes Γ_n and Γ_b and setting

$$\mathcal{E} = \Phi_0 \varepsilon^{-1/4} (\Gamma_n \mathbf{n} + \Gamma_b \mathbf{b}) \exp\left(ik_0 \int n\, ds\right). \tag{2.36}$$

instead of (2.26). Henceforward we shall omit the index of the zeroth approximation "0" to make formulae more readable. Here Φ_0 is a real-valued function that satisfies the law of energy conservation along the ray tube

$$\operatorname{div}(\Phi_0^2 \mathbf{t}) = 0 \tag{2.37}$$

(Born and Wolf, 1980; Kravtsov and Orlov, 1990). Normalized amplitudes have unit intensity: $|\Gamma_n|^2 + |\Gamma_b|^2 = 1$.

In designations (2.36), Eqs. (2.35) can be rewritten as

$$\frac{d\Gamma_n}{ds} - \frac{\Gamma_b}{T} = 0, \qquad \frac{d\Gamma_b}{ds} + \frac{\Gamma_n}{T} = 0. \tag{2.38}$$

For the angle θ between the vector $\mathbf{E}$ and the unit vector $\mathbf{n}$, (Figure. 2.2)

$$\theta = \arctan \frac{E_b}{E_n} = \arctan \frac{\Gamma_b}{\Gamma_n}, \tag{2.38a}$$

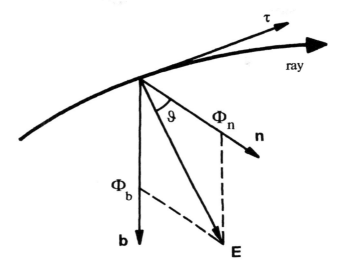

FIGURE 2.2. The position of the electric intensity vector **E** relatively to the normal **n**, binormal **b** and tangent τ to the ray.

the well-known Rytov equation (Rytov, 1938) follows from Eq. (2.38):

$$\frac{d\theta}{ds} = -\frac{1}{T}. \tag{2.39}$$

System of Eqs. (2.38), along with the pair of Eqs. (2.37) and (2.39), fully defines the zero-order field in an isotropic medium. The Rytov law of polarization rotation is a particular case of a more general law which is referred to as the Berry phase (Berry, 1984). These are reviewed in a recent work by Klyshko (1993).

2.3   Geometrical optics of an anisotropic medium (the Courant–Lax method)

This method, partially discussed in Section 2.1, allows for a correct transition to the case of a homogeneous anisotropic medium for which the coefficients C_1 and C_2 in Eq. (2.15) are constant. Thus, it implicitly assumes that the normal waves are non-interacting.

In the theory by Courant and Lax an ordinary differential equation (the transfer equation), localized along rays (2.10), is derived for each of the scalar amplitudes C_1 and C_2. For the extraordinary ray (i. e. with respect to C_1), for example, this can be done by setting in Eqs. (2.30) and (2.32)

$$\tilde{\mathbf{q}}_1 = \mathbf{e}_{1\perp}(\mathbf{r}; \mathbf{k}_1/|\mathbf{k}_1|) = e_{11}\mathbf{q}_1 + e_{12}\mathbf{q}_2,$$

$$\tilde{\mathbf{q}}_2 = \mathbf{e}_{2\perp}(\mathbf{r}; \mathbf{k}_1/|\mathbf{k}_1|) = e_{21}\mathbf{q}_1 + e_{22}\mathbf{q}_2, \qquad (2.40)$$

$$\mathbf{Q}_1 = \mathbf{e}_1^*(\mathbf{r}; \mathbf{k}_1/|\mathbf{k}_1|), \mathbf{Q}_2 = \mathbf{e}_2^*(\mathbf{r}; \mathbf{k}_1/|\mathbf{k}_1|),$$

where the coefficients $e_{11}, e_{12}, e_{21}, e_{22}$ and vectors $\mathbf{q}_1, \mathbf{q}_2, \mathbf{Q}_1, \mathbf{Q}_2$ are related to the vector $\mathbf{k}_1$ by formulae (2.17)–(2.20). $e_{1\perp}$ and $e_{2\perp}$ here denote the components of polarization unit vectors $\mathbf{e}_1$ and $\mathbf{e}_2$ that are transverse to $\mathbf{k}_1$. Thus, the sought-for amplitude C_1 is connected with the amplitude D_1 which enters (2.32) by the relationship

$$C_1 = n_1^{-2}D_1. \qquad (2.41)$$

For the wave of a given (first) type, the amplitude D_2 describes a polarization correction due to inhomogeneity of the medium which leaves the phase and ray structure of the wave completely unperturbed. In the zeroth order of the Courant–Lax method this amplitude, D_2, is assumed to be equal to zero. Therefore, in this method one is concerned with a single equation for a single unknown D_1 (or C_1), i. e. with a wave constructed in accordance with (2.15), instead of constructing the system of two equations along a ray (as in the Rytov method). For a wave of the first type, the first of the universal equations (2.30) containing the only variable D_1 takes the "ray" form mentioned. The second of them then yields D_2 (or C_2) in the next order in terms of D_1 or C_1.

Indeed, in accord with (2.40) the first of equations (2.30) can be rewritten as

$$(\mathbf{h}_1^* \times \mathbf{e}_1 - \mathbf{e}_1^* \times \mathbf{h}_1)\nabla C_1 + (-\mathbf{h}_1^* \operatorname{curl} \mathbf{e}_1 + \mathbf{e}_1 \operatorname{curl} \mathbf{h}_1)C_1 = 0, \quad (2.42)$$

where $\mathbf{h}_1 = k_0^{-1}\mathbf{k}_1 \times \mathbf{e}_1$ is the magnetic polarization vector. On the other hand, by replacing $\mathbf{k} \to \mathbf{k}_1, \mathbf{E}^{(0)} \to \mathbf{e}_1$ in Eq. (2.4) one readily obtains the identity

$$\mathbf{v}_{g1} = \left(\frac{\partial \Omega_1}{\partial \mathbf{k}}\right)\Bigg|_{\mathbf{k}=\mathbf{k}_1} = \frac{c(\mathbf{e}_1^* \times \mathbf{h}_1 - \mathbf{h}_1^* \times \mathbf{e}_1)}{2(\mathbf{e}_1^* \mathbf{d}_1)}. \qquad (2.43)$$

Correlating this with the Hamiltonian system of equations (2.10), we rearrange (2.42) in the form

$$\frac{dC_1}{dt_1} + P_1 C_1 = 0. \qquad (2.42a)$$

Here d/dt_1 implies differentiation with respect to time t along the extraordinary ray described by the first system of Hamiltonian equations (2.10), whereas

$$P_1 = \frac{c(\mathbf{h}_1^* \operatorname{curl} \mathbf{e}_1 - \mathbf{e}_1^* \operatorname{curl} \mathbf{h}_1)}{2(\mathbf{e}_1^* \mathbf{d}_1)}. \qquad (2.44)$$

The possibility of grouping partial derivatives in a full derivative along a ray is not a specific feature of Maxwell's equations. As shown by Courant and Lax (1956) a similar grouping is typical of any linear system of wave type equations while the important relationship (2.43) is a particular case of the so-called *Courant–Lax fundamental identity*. The choice of vectors $\mathbf{e}_1^*$ and $\mathbf{e}_2^*$ for $\mathbf{Q}_1$ and $\mathbf{Q}_2$ is also not arbitrary: as is implicit in the Courant–Lax method, $\mathbf{e}_1^*$ and $\mathbf{e}_2^*$ are left eigenvectors of the matrix of vector equation (2.4) (in which the substitution $\mathbf{k} \to \mathbf{k}_1$ is made). This implies that $\mathbf{e}_1^*$ and $\mathbf{e}_2^*$ ensure the identities

$$\mathbf{e}_1^*(k_0^2 \hat{\varepsilon} \mathbf{A} + \mathbf{k}_1 \times \mathbf{k}_1 \times \mathbf{A}) \equiv 0, \qquad (2.45)$$

$$\mathbf{e}_2^*(k_0^2 \hat{\varepsilon} \mathbf{A} + \mathbf{k}_1 \times \mathbf{k}_1 \times \mathbf{A}) \equiv k_0^2(1 - n_1^2 n_2^{-2}) \mathbf{e}_2^* \hat{\varepsilon} \mathbf{A}, \qquad (2.46)$$

with the vector $\mathbf{A}$ being arbitrary.

The next, after (2.15), order of the perturbation theory could be derived by substituting Eqs. (2.32) in a full form, in conjunction with (2.40), into the second of Eqs.(2.30). This yields, in accord with (2.4), (2.46) and (2.22), the expression for C_2 in first order of perturbation theory:

$$C_2 = ik_0^{-2} n_2^2 (n_2^2 - n_1^2)^{-1} \mathbf{e}_2^*(\mathbf{k}_1 \times \operatorname{curl}(C_1 \mathbf{e}_1) + \operatorname{curl}(\mathbf{k}_1 \times C_1 \mathbf{e}_1)). \qquad (2.47)$$

In a similar way a complete recurrent series could be generated in all subsequent orders of the perturbation theory.

One readily sees that the correction (2.47) diverges as one approaches an isotropic medium where $n_1 = n_2$. This implies that the Courant–Lax method no longer applies (it diverges) as $\delta \lesssim 1$ and is not capable of solving the problem of limit polarization.

Notice that the Courant–Lax method also contains an "isotropic" branch which satisfies the Rytov equations (2.35). However, this method did not envisage a modification that would be applicable in the case of weak anisotropy.

2.4 Geometrical optics of weakly anisotropic media: the quasi-isotropic approximation (QIA)

2.4.1 BASIC ASSUMPTIONS OF THE QIA.

The quasi-isotropic approximation of geometrical optics is based on the choice of the zero-approximation solution $\mathcal{E}^{(0)}$ in the form (2.26), as if there were no anisotropy (Kravtsov, 1968a). In electrodynamics this conforms to the case of the zero anisotropy tensor $\nu_{mn} = \varepsilon_{mn} - \varepsilon_o \delta_{mn}$. In the framework of the quasi-isotropic approach the zeroth approximation can conveniently be presented in the "isotropic" form (2.36) where a replacement $\varepsilon(\mathbf{r}) \to \varepsilon_0(\mathbf{r})$ is to be made, with the choice of scalar quantity ε_0 being arbitrary to a certain extent. One may take one of the diagonal components of $\widehat{\varepsilon}$ or its mean value $\varepsilon_0 = \frac{1}{3} \operatorname{Tr} \widehat{\varepsilon}$, as well as any similar quantity, say, $\varepsilon_0 = (\det \widehat{\varepsilon})^{1/3}$. There exists only one but necessary constraint which is that components of the anisotropy tensor ν_{mn} should be small when compared to ε_0.

Then, we apply the isotropic eikonal formula (2.36) to describe waves in a weakly anisotropic medium. The quantity ds there denotes an element of ray length which corresponds to an isotropic medium with a dielectric permittivity $\varepsilon_0(\mathbf{r})$. The unit vectors $\mathbf{n}, \mathbf{b}$, and $\mathbf{t}$ are the normal, binormal, and tangent vectors to the "isotropic" ray. The ray obeys the Hamiltonian equations (2.12) in which a substitution of $n = n_0 = \varepsilon_0^{1/2}$ has to be made. Figure 1.1 shows the "isotropic" ray by a dashed line. The factor Φ_0 in (2.36) obeys the conservation law (2.37).

As yet unknown normalized amplitudes Γ_n and Γ_b can be obtained from (2.30) projected onto the unit vectors $\mathbf{Q}_1 = \mathbf{n}$ and $\mathbf{Q}_2 = \mathbf{b}$. In addition to the isotropic case we should make allowance for terms containing the anisotropy tensor ν_{mn} assuming them to scale as μ and taking them into account in a first-order equation in the small parameter μ. It is the fact that small anisotropic terms are taken into account that is the essence of the QIA.

2.4.2 "Standard" form of QIA equations

Applying the universal procedure (2.30) with "isotropic" unit vectors $\mathbf{Q}_1 = \mathbf{n}$ and $\mathbf{Q}_2 = \mathbf{b}$ and using the Rytov identity (2.33) we are basically using the same operations as in the Rytov method (Section 2.2), the only difference being in the appearance of terms characterizing the weak medium anisotropy. As a result the two relations (2.30) reduce to the form (Kravtsov, 1968a):

$$\frac{d\Gamma_n}{ds} - \frac{1}{2}ik_0\varepsilon_0^{-1/2}(\nu_{nn}\Gamma_n + \nu_{nb}\Gamma_b) - T^{-1}\Gamma_b = 0,$$

$$\frac{d\Gamma_b}{ds} - \frac{1}{2}ik_0\varepsilon_0^{-1/2}(\nu_{bn}\Gamma_n + \nu_{bb}\Gamma_b) + T^{-1}\Gamma_n = 0. \tag{2.48}$$

Here indices n and b refer, respectively, to the normal $\mathbf{n}$ and binormal $\mathbf{b}$ to the ray, whereas quantities

$$\nu_{nn} = (\mathbf{n}, \widehat{\nu}\mathbf{n}), \ \nu_{nb} = (\mathbf{n}, \widehat{\nu}\mathbf{b}), \ \nu_{bn} = (\mathbf{b}, \widehat{\nu}\mathbf{n}), \ \nu_{bb} = (\mathbf{b}, \widehat{\nu}\mathbf{b})$$

are the components of the anisotropy tensor written in the system of unit vectors $\mathbf{n}, \mathbf{b}$, and $\mathbf{t}$ associated with the rays.

System (2.48) will be referred to as the "standard" form of QIA equations. From this system a very convenient equation follows for the polarization parameter $\theta = \arctan \Gamma_b/\Gamma_n$. We will consider it in Section 2.4.3. A small modification will bring us to very similar but more accurate QIA equations for the induction vector in Section 3.1 and to QIA equations in the form of interacting normal waves in Section 3.2.

In Section 3.3 we reject the hypothesis that both rays, ordinary and extraordinary, practically coincide with an isotropic ray, and suggest a modification of the QIA in the form of a split ray method admitting spatial separation of ordinary and extraordinary rays. In Section 3.4 we also lift the constraint of weak anisotropy and propose a synthetic approach suited not only for weakly anisotropic media, but also for strongly anisotropic media.

In Chapter 4 the QIA equations will be applied to analyze the propagation of electromagnetic waves in a plasma, while in Chapter 5 they will be used for the analysis of optical phenomena. Then the QIA viewpoint will be extended to wave problems arising in elasticity theory (Chapter 6) and to spinor wave functions (Chapter 7).

2.4.3   QIA EQUATIONS FOR COMPLEX POLARIZATION PARAMETER

Introducing the angle of polarization θ (which is complex, generally speaking) by formula (2.38a),

$$\theta = \arctan \frac{E_b}{E_n} = \arctan \frac{\Gamma_b}{\Gamma_n},$$

we may turn from Eqs. (2.48) to a single equation for the parameter θ.
Indeed, by virtue of (2.48)

$$\frac{d\theta}{ds} = \frac{d}{ds}\arctan\frac{\Gamma_b}{\Gamma_n} = \frac{\Gamma_n \dfrac{d\Gamma_b}{ds} - \Gamma_b \dfrac{d\Gamma_n}{ds}}{\Gamma_n^2 + \Gamma_b^2}$$

$$= \frac{1}{\Gamma_n^2 + \Gamma_b^2}\left[-T^{-1}(\Gamma_n^2 + \Gamma_b^2) + iM(\nu_{bb} - \nu_{nn})\Gamma_n\Gamma_b \right. \tag{2.49}$$

$$\left. +iM(\nu_{bn}\Gamma_n^2 - \nu_{nb}\Gamma_b^2)\right],$$

where we temporarily set $M = (1/2)k_0\varepsilon_0^{-1/2}$.
Eq. (2.49) can be rearranged as

$$\frac{d\theta}{ds} + T^{-1} + \frac{i}{2}M(\nu_{nb} - \nu_{bn})$$

$$+iM(\nu_{nn} - \nu_{bb})\frac{\Gamma_n\Gamma_b}{\Gamma_n^2 + \Gamma_b^2} - \frac{iM}{2}(\nu_{nb} + \nu_{bn})\frac{\Gamma_n^2 - \Gamma_b^2}{\Gamma_n^2 + \Gamma_b^2} = 0. \tag{2.50}$$

According to (2.38a)

$$\frac{\Gamma_n\Gamma_b}{\Gamma_n^2 + \Gamma_b^2} = \frac{\tan\theta}{1 + \tan^2\theta} = \sin\theta\cos\theta = \frac{1}{2}\sin 2\theta,$$

$$\frac{\Gamma_n^2 - \Gamma_b^2}{\Gamma_n^2 + \Gamma_b^2} = \frac{1 - \tan^2\theta}{1 + \tan^2\theta} = \cos^2\theta - \sin^2\theta = \cos 2\theta.$$

As a result we conclude that θ obeys the equation (Kravtsov, 1968a)

$$\frac{d\theta}{ds} + [T^{-1} + \frac{1}{4}ik_0\varepsilon_0^{-1/2}(\nu_{nb} - \nu_{bn})]$$

$$+\frac{1}{4}ik_0\varepsilon_0^{-1/2}[(\nu_{nn} - \nu_{bb})\sin 2\theta - (\nu_{nb} + \nu_{bn})\cos 2\theta] = 0. \tag{2.51}$$

This equation is a variant of the Riccati equation. At $\hat{\nu} = 0$ (an isotropic medium) it reduces to the Rytov equation (2.39).

The complex angle θ characterizes all parameters of the polarization ellipse. Its real part $\theta' = \text{Re}\,\theta$ gives the inclination of the major ellipse axis measured from the normal $\mathbf{n}$ to the ray. The hyperbolic tangent of the imaginary part $\theta'' = \text{Im}\,\theta$ equals the ratio of small axis b to the large axis, a: $|\tanh\theta''| = b/a$, while the sign of θ'' characterizes the sense of rotation of field vectors: if $\theta'' > 0$ the vector rotates clockwise, and if $\theta'' < 0$ it rotates counterclockwise viewed along the ray.

Were Eq. (2.51) for the complex polarization angle θ to be solved in one way or another, the complex amplitude

$$\Gamma = (\Gamma_n^2 + \Gamma_b^2)^{1/2} \tag{2.52}$$

could then be expressed by a simple quadrature. In accord with (2.48),

$$\frac{d\Gamma^2}{ds} = 2\Gamma_n \frac{d\Gamma_n}{ds} + 2\Gamma_b \frac{d\Gamma_b}{ds}$$

$$= 2iM\left[\nu_{nn}\Gamma_n^2 + \nu_{bb}\Gamma_b^2 + (\nu_{nb} + \nu_{bn})\Gamma_n\Gamma_b\right], \tag{2.53}$$

where $M = (1/2)k_0\varepsilon_0^{-1/2}$ is the same parameter as that entering (2.49) and (2.50).

Dividing both sides of (2.53) by $2\Gamma^2 = 2(\Gamma_n^2 + \Gamma_b^2)$ we obtain

$$\frac{1}{2\Gamma^2}\frac{d\Gamma^2}{ds} = \frac{d\ln\Gamma}{ds} = \frac{iM}{2}\left[(\nu_{nn} + \nu_{bb})\right.$$

$$\left. +(\nu_{nn} - \nu_{bb})\cos 2\theta + (\nu_{nb} + \nu_{bn})\sin 2\theta\right]. \tag{2.54}$$

By virtue of (2.54) the amplitude Γ is expressed through θ by quadrature

$$\Gamma = \exp\left\{\frac{iM}{2}\left[(\nu_{nn} + \nu_{bb})\right.\right.$$

$$\left.\left. +(\nu_{nn} - \nu_{bb})\cos 2\theta + (\nu_{nb} + \nu_{bn})\sin 2\theta\right]\right\}. \tag{2.55}$$

On obtaining θ from (2.51) and Γ from (2.55), the components Γ_n and Γ_b can be found from the obvious relations:

$$\Gamma_n = \Gamma\cos\theta, \quad \Gamma_b = \Gamma\sin\theta, \tag{2.56}$$

valid for either real or complex values of the complex polarization parameter θ.

It is significant that the complex amplitude $\Gamma = (\Gamma_n^2 + \Gamma_b^2)^{1/2}$ (2.52) should be distinguished from the real amplitude

$$|\Gamma| = 2(|\Gamma_n^2| + |\Gamma_b^2|)^{1/2}. \qquad (2.57)$$

The distinction between them is most pronounced with a circular polarization, for which $\Gamma_b = \pm\Gamma_n$. In this case

$$|\Gamma|^2 = |\Gamma_n|^2 + |\Gamma_b|^2 = 2|\Gamma_n|^2,$$

whereas

$$\Gamma^2 = \Gamma_n^2 + \Gamma_b^2 = \Gamma_n^2 - \Gamma_n^2 = 0.$$

Needless to say, an equation for $|\Gamma|^2$ differs from that for Γ. It is easy to obtain from (2.48) that

$$\frac{d|\Gamma|^2}{ds} = \Gamma_n^* \frac{d\Gamma_n}{ds} + \Gamma_n \frac{d\Gamma_n^*}{ds} + \Gamma_b^* \frac{d\Gamma_b}{ds} + \Gamma_b \frac{d\Gamma_b^*}{ds}$$

$$= 2iM \left[\nu_{nn} |\Gamma_n|^2 + \nu_{bb} |\Gamma_b|^2 + \nu_{nb} \Gamma_n^* \Gamma_b + \nu_{bn} \Gamma_n \Gamma_b^* - \text{c.c} \right], \qquad (2.58)$$

where c.c denotes complex conjugate terms.

With the notations

$$\nu_{mn}^h = \frac{1}{2}(\nu_{mn} + \nu_{mn}^*)$$

and

$$\nu_{mn}^a = \frac{1}{2}(\nu_{mn} - \nu_{mn}^*),$$

respectively, for Hermitian and anti-Hermitian parts of the tensor ν_{mn} (2.58) becomes

$$\frac{d|\Gamma|^2}{ds} = -2M \left[\nu_{nn}^a |\Gamma_n|^2 + \nu_{bb}^a |\Gamma_b|^2 + \nu_{nb}^a \Gamma_n \Gamma_b + \nu_{bn}^a \Gamma_n \Gamma_b \right]. \qquad (2.59)$$

According to (2.59) the quantity $|\Gamma|$ remains constant only in non-absorbing media in which the anti-Hermitian part of the anisotropy tensor equals zero:

$$\nu_{\alpha\beta}^a = 0.$$

It is noteworthy that the absorption in a medium changes not only the energy of a propagating wave, but also its polarization. We refer to the phenomenon of dichroism (or pleochroism) arising in media with unlike coefficients of normal mode absorption as the best known effect of that type. That the absorption coefficients in such media are not identical results in the survival of only one of the two normal waves at large distances. Accordingly, limiting polarization coincides with that of the normal wave which survives and turns to be different from the initial polarization which is determined by the superposition of two normal waves.

2.4.4 THE QIA EQUATIONS IN MOVING AXES.

In many particular cases, the eikonal formula (2.24) and Eqs. (2.29) can be conveniently written in moving axes $\mathbf{q}_1(s)$ and $\mathbf{q}_2(s)$ which satisfy the condition of orthogonality $\mathbf{q}_1 \perp \mathbf{q}_2 \perp \mathbf{t}$. For example, the eigenaxes of two-dimensional tensor with components

$$\mathrm{Re} \begin{pmatrix} \varepsilon_{nn} & \varepsilon_{nb} \\ \varepsilon_{bn} & \varepsilon_{bb} \end{pmatrix} \tag{2.60}$$

can be used for $\mathbf{q}_1$ and $\mathbf{q}_2$ axes.

We set

$$\mathbf{q}_1 = \mathbf{n} \cos \psi + \mathbf{b} \sin \psi, \quad \mathbf{q}_2 = -\mathbf{n} \sin \psi + \mathbf{b} \cos \psi. \tag{2.60a}$$

Then the substitution

$$\mathcal{E} = \mathbf{E} \exp \left(ik_0 \int n_0 \, ds \right), \quad \mathbf{E} = \Phi_0 \varepsilon_0^{-1/4} (\Gamma_1 \mathbf{q}_1 + \Gamma_2 \mathbf{q}_2) \tag{2.61}$$

is used instead of (2.36) whereas Eqs. (2.48) takes the form

$$\frac{d\Gamma_1}{ds} - \frac{1}{2} ik_0 \varepsilon_0^{-1/2} (\nu_{11}\Gamma_1 + \nu_{12}\Gamma_2) - T_{\text{eff}}^{-1} \Gamma_2 = 0,$$

$$\frac{d\Gamma_2}{ds} - \frac{1}{2} ik_0 \varepsilon_0^{-1/2} (\nu_{21}\Gamma_1 + \nu_{22}\Gamma_2) + T_{\text{eff}}^{-1} \Gamma_1 = 0. \tag{2.62}$$

Indices 1, 2 here correspond to unit vectors $\mathbf{q}_1$ and $\mathbf{q}_2$:

$$\nu_{\alpha\beta} = (\mathbf{q}_\alpha, \hat{\nu}\mathbf{q}_\beta) \quad (\alpha, \beta = 1, 2).$$

The value

$$T_{\text{eff}}^{-1} = T^{-1} + d\psi/ds \tag{2.63}$$

represents an effective ray torsion in moving axes $\mathbf{q}_1$ and $\mathbf{q}_2$; ψ is the angle of rotation from the normal $\mathbf{n}$ to the unit vector $\mathbf{q}_1$:

$$\psi = \arctan\left(\mathbf{b}\mathbf{q}_1/\mathbf{n}\mathbf{q}_1\right). \tag{2.64}$$

2.4.5 "INVARIANT" FORM OF THE QIA

Sometimes it is convenient to write (2.61) and (2.62) in the "invariant" form

$$\mathcal{E} = \tilde{\mathbf{E}} \exp\left(\frac{1}{4}ik_0 \int \varepsilon_0^{-1/2}(2\varepsilon_0 + \varepsilon_{11} + \varepsilon_{22})\,ds\right),$$

$$\tilde{\mathbf{E}} = \Phi_0 \varepsilon_0^{-1/4}(\tilde{\Gamma}_1 \mathbf{q}_1 + \tilde{\Gamma}_2 \mathbf{q}_2), \tag{2.65}$$

$$\frac{d\tilde{\Gamma}_1}{ds} - \frac{1}{2}ik_0\varepsilon_0^{-1/2}(\gamma_{11}\tilde{\Gamma}_1 + \gamma_{12}\tilde{\Gamma}_2) - T_{\text{eff}}^{-1}\tilde{\Gamma}_2 = 0,$$

$$\frac{d\tilde{\Gamma}_2}{ds} - \frac{1}{2}ik_0\varepsilon_0^{-1/2}(\gamma_{21}\tilde{\Gamma}_1 + \gamma_{22}\tilde{\Gamma}_2) + T_{\text{eff}}^{-1}\tilde{\Gamma}_1 = 0, \tag{2.66}$$

where

$$\gamma_{11} = -\gamma_{22} = (\nu_{11} - \nu_{22})/2, \quad \gamma_{12} = \nu_{12}, \quad \gamma_{21} = \nu_{21}. \tag{2.66a}$$

In particular, the unit vectors $\mathbf{n}$ and $\mathbf{b}$ might be taken as the unit vectors $\mathbf{q}_1$ and $\mathbf{q}_2$ in (2.65) and (2.66). To make the transformation to these unit vectors one needs to change tensor indices: $1 \to n, 2 \to b$.

One can easily see that the argument in the exponent in (2.65) is independent of variations of the scalar $\varepsilon_0(\mathbf{r})$ to first order in the small parameter μ_1. The same invariance is exhibited by the coefficients $\varepsilon_0^{-1/2}\gamma_{\alpha\beta}$ of Eqs. (2.66). Hence, the quasi-isotropic approximation in the form (2.65), (2.66) is invariant to first order in μ_1 with respect to the mentioned variations of the function $\varepsilon_0(\mathbf{r})$.

Particular methods of solving Eqs.(2.48) and (2.66) will be presented in Sections 2.5 – 2.7, while in subsequent sections they will be applied to the analysis of separate physical effects. In this section we restrict ourselves to the illustration of the possibility of an approximate transition from Eqs. (2.66) to the independent normal waves in a homogeneous medium.

2.4.6 PASSING FROM THE QIA TO "SIMPLIFIED" NORMAL WAVES.

In a homogeneous medium ($T^{-1} = 0, \varepsilon_{\alpha\beta} = $ const) Eqs. (2.65) and (2.66) lead to the following expressions for the normal waves:

$$\mathcal{E}_{1,2} = \tilde{e}_{1,2} \exp\left(ik_0 \tilde{n}_{1,2} s\right), \tag{2.67}$$

where we denoted

$$\tilde{n}_{1,2} = \frac{1}{4}\varepsilon_0^{-1/2} \left[2\varepsilon_0 + \varepsilon_{11} + \varepsilon_{22} \mp \left((\varepsilon_{11} - \varepsilon_{22})^2 + 4|\varepsilon_{12}|^2\right)^{1/2}\right],$$

$$\tilde{\mathbf{e}}_1 = \tilde{e}_{11}\mathbf{q}_1 + \tilde{e}_{12}\mathbf{q}_2,$$

$$\tilde{\mathbf{e}}_2 = \tilde{e}_{21}\mathbf{q}_1 + \tilde{e}_{22}\mathbf{q}_2, \tag{2.68}$$

$$\tilde{e}_{11} = \tilde{e}_{22} = (1 + \tilde{K}_1^2)^{-1/2}, \quad \tilde{e}_{12} = \tilde{e}_{21} = -i\tilde{K}_1(1 + \tilde{K}_1^2)^{-1/2},$$

$$\tilde{K}_1 = J\tilde{K}^J, \quad J = \operatorname{sgn} \operatorname{Im}\varepsilon_{12}, \quad \tilde{K} = \tilde{Q} - (1 + \tilde{Q}^2)^{1/2},$$

$$\tilde{Q} = i(\varepsilon_{22} - \varepsilon_{11})/(2\varepsilon_{12}).$$

It can readily be seen that to first order in the small perturbation $\nu_{\alpha\beta}$ the eikonals of expressions (2.67) coincide with the eikonals $k_0 n_1 s$ and $k_0 n_2 s$ of the extraordinary and ordinary waves, in which n_1 and n_2 are given by formula (2.7) after replacing the lower indices $x \to 1, y \to 2$.

The difference in refractive indices, $\Delta n = |n_1 - n_2|$, according to (2.68) is

$$\Delta n = |n_1 - n_2| = \varepsilon_0^{-1/2} \left[\frac{(\nu_{11} - \nu_{22})^2}{4} + |\nu_{12}|^2\right]^{1/2}, \tag{2.69}$$

i.e., it is scaled as $\mu_1 = \max|\nu_{mn}|$.

Similarly, expressions for $\tilde{e}_1$ and $\tilde{e}_2$ in (2.68) coincide with expressions for $e_{1\perp}$ and $e_{2\perp}$ that follow from formulae (2.18) -(2.20) to first order in $\nu_{\alpha\beta}$. Also, all formulae (2.68) which express the components

$\tilde{e}_{11}, \tilde{e}_{12}, \tilde{e}_{21}, \tilde{e}_{22}$ through the ratio $(\varepsilon_{11} - \varepsilon_{22})/\varepsilon_{12}$ are completely identical to formulae (2.19), (2.20) expressing the components $e_{11}, e_{12}, e_{21}, e_{22}$ through the ratio $(\chi_{11} - \chi_{22})/\chi_{12}$.

Therefore expressions (2.67) and (2.68) can be termed as "simplified" normal waves. They coincide with exact normal waves up to the terms of μ_1 order both in the argument of the exponent and in the amplitude factors.

It seems clear that the difference between the "simplified" and exact eikonals will grow with components $|\nu_{13}|$ and $|\nu_{23}|$ of the anisotropy tensor because of terms of μ_1^2 or higher orders and may become noticeable if the condition

$$\max\left(|\nu_{13}|^2, |\nu_{23}|^2\right) \lesssim \max\left(|\nu_{11}|, |\nu_{12}|\right) \tag{2.70}$$

is violated. Corresponding differences will also appear in amplitudes. In this way the inequalities (2.70) serve as the applicability condition for Eqs. (2.65) and (2.66) and formulas that follow from them.

The presentation of the wave field as a superposition of independent (though "simplified") normal waves is feasible subject to the inequality $\delta \gg 1$, that is, under weak interaction. For $\delta \sim 1$, and the more so, $\delta < 1$ the wave interaction cannot be discarded. In this case the QIA equations admit a rearrangement into a system of two coupled equations for amplitudes of *interacting* normal waves. We write them in Section 3.2. In that section we also compare the QIA with Budden's method for a plane-layered medium.

2.5   Methods of solution of the QIA equations

2.5.1   INTRODUCTORY REMARKS

QIA equations form the system of two coupled ordinary differential equations with variable coefficients, i.e., a system like

$$\frac{d\,x(t)}{d\,t} = a_{11}(t)x(t) + a_{12}(t)y(t),$$

$$\frac{d\,y(t)}{d\,t} = a_{21}(t)x(t) + a_{22}(t)y(t). \tag{2.71}$$

Such equations are encountered in many fields of physics being applied to the description of a number of similar phenomena: adiabatic perturbations in non-stationary problems of quantum mechanics (Landau and Lifshitz, 1977; Sobelman, 1992; Sobelman, Vainstein and Yukov, 1995) and the phenomenon of spin reversal in beams of polarized particles, linear interaction of normal oscillations in nonstationary systems (Erokhin and Moiseev, 1973), the mutual transformation of normal modes in inhomogeneous waveguides.

Special methods developed when solving the problems mentioned could be helpful also for solution of QIA equations. These methods include:

i) *Numerical methods* in which respect we restrict ourselves only to general comments (Section 2.5.2).

ii) *The method of small perturbations* which applies provided the coefficients a_{ij} vary slightly: $|\Delta a_{ik}| \ll |a_{ik}|$. A few results obtained by this method will be noted below, in Section 2.5.3.

iii) *Exact asymptotic solutions*. The exact solutions to the system of coupled equations (2.71) are known for a very limited number of particular cases. Specifically, a solution of (2.71) in the form of normal harmonic oscillations corresponds to constant coefficients a_{ij}, whereas the case of linearly varying coefficients is settled by the Landau–Zener solution. We describe the latter in Section 2.5.4 and then apply it in Section 4.2 to the analysis of wave transformation in a magnetized plasma.

iv) *Asymptotic methods of solution*. Asymptotic approaches to the solution of coupled equations (2.71) are discussed shortly in Section 2.5.5.

2.5.2   NUMERICAL METHODS

In the context of practical problems the QIA equations can be solved by standard numerical methods, for example, by the Runge-Kutta method. Apparently, one does not need to use numerical calculations along the whole ray. Sometimes it suffices to perform numerical analysis only in the regions of small, $\delta \ll 1$, and moderate, $\delta \lesssim 1$, birefringence. In the region of relatively strong birefringence, $\delta \gg 1$, direct numerical methods are inefficient due to fast oscillations of the wave field. On the other hand, in that region an iterative approach suggested by Naida (1977a, 1978b) corresponding to the method of split rays (see Section 3.3 below) is especially efficient.

2.5.3   PERTURBATION METHOD

If the size of the region of weak birefringence, $\delta \ll 1$, is large enough, numerical methods may appear to be inappropriate. However, compensating this, Eqs.(2.48) can be solved in such a region by a perturbation method in a small parameter $\delta = k_0 l |n_1 - n_2|$. Indeed, if the initial values Γ_n^{in} and Γ_b^{in} are specified at some point s_{in} of the ray, for example, calculated by the Runge-Kutta method on leaving the region of moderate birefringence $\delta \leq 1$, then for the points in the interaction region we find

$$\Gamma_n(s) = \int_{s_{in}}^{s} \left[\frac{1}{2} i k_0 \varepsilon_0^{-1/2} (\nu_{nn} \Gamma_n^{in} + \nu_{nb} \Gamma_b^{in}) + T^{-1} \Gamma_b^{in} \right] ds,$$

$$\Gamma_b(s) = \int_{s_n}^{s} \left[\frac{1}{2} i k_0 \varepsilon_0^{-1/2} (\nu_{nb} \Gamma_n^{in} + \nu_{bb} \Gamma_b^{in}) - T^{-1} \Gamma_n^{in} \right] ds. \tag{2.72}$$

If the contribution from the torsion T^{-1} considerably exceeds the contribution from the coefficient of anisotropy, (2.72) may be rewritten as

$$\Gamma_n(s) = \Gamma_n^0(s) + \int_{s_{in}}^{s} \left[\frac{1}{2} i k_0 \varepsilon_0^{-1/2} (\nu_{nn} \Gamma_n^0 + \nu_{nb} \Gamma_b^0) \right] ds,$$

$$\Gamma_b(s) = \Gamma_b^0(s) + \int_{s_n}^{s} \left[\frac{1}{2} i k_0 \varepsilon_0^{-1/2} (\nu_{nb} \Gamma_n^0 + \nu_{bb} \Gamma_b^0) \right] ds. \tag{2.73}$$

Here Γ_n^0 and Γ_b^0 are the solutions to Eqs.(2.38) subject to the initial conditions specified above, i. e. accounting only for the ray torsion:

$$\Gamma_n^0 = \Gamma_n^{in} \cos \int_{s_{in}}^{s} T^{-1} ds + \Gamma_b^{in} \sin \int_{s_{in}}^{s} T^{-1} ds,$$

$$\Gamma_b^0 = -\Gamma_n^{in} \sin \int_{s_{in}}^{s} T^{-1} ds + \Gamma_b^{in} \cos \int_{s_{in}}^{s} T^{-1} ds.$$

Note that solutions of (2.72) and (2.73) type can also be derived from the QIA equations in the form of a Riccati equation (2.51) (Naida 1970, 1971). Another example of the application of perturbation theory to the QIA equations will be given in Section 4.2.5 of Chapter 4.

Solutions of QIA equations by the perturbation method yields practically the same results as it would if the perturbation theory were applied to the initial wave problem (see, for example, Ginzburg, 1970; Kucherenko, 1974). The distinction is perhaps apparent only in the ability of perturbation theory applied to the full, original equations to describe the wave reflection from the stepwise change in parameters of the medium, in particular, from the region where the components of the anisotropy tensor and/or its derivatives suffer a discontinuity. In the QIA framework the perturbation theory is constrained by the requirement of smooth variations in parameters of the medium and is therefore not capable in principle of describing the wave reflection. On the other hand, the perturbation theory in the QIA framework is simpler and allows for a more intuitive interpretation of results.

2.5.4   METHOD OF LINEARIZATION OF COEFFICIENTS

In many practical problems concerning a localized interaction the size of the moderate interaction region, $\delta \lesssim 1$, is sufficiently small. Within this region the coefficients of the QIA equations, for instance, those of Eq. (2.48), (2.62), or (2.66), may be approximated by linear functions of the ray length s.

Assume the tensor ν_{mn} to be Hermitian, $\nu_{nm} = \nu_{mn}^*$, and, for ε_0, take $\varepsilon_0 = (1/2)(\varepsilon_{11} + \varepsilon_{22})$, so that $\nu_{11} = -\nu_{22} = (1/2)(\varepsilon_{11} - \varepsilon_{22})$. Also, to simplify the analysis we assume $T^{-1} = 0$, i.e. the ray torsion will be ignored. A linear approximation for the coefficients $\nu_{11} = -\nu_{22}$ and $\nu_{12} = \nu_{21}^*$ in the vicinity of a given point s_0 takes the form

$$\nu_{11} = -\nu_{22} \approx \nu_{11}^0 + v(s - s_0),$$
$$\nu_{12} = \nu_{21}^* \approx \nu_{12}^0 + V(s - s_0),$$
(2.74)

where

$$v = \frac{\partial \nu_{11}(s_0)}{\partial s}, \quad V = \frac{\partial \nu_{12}(s_0)}{\partial s}.$$
(2.75)

On the order of magnitude, the coefficients v and V are inversely proportional to l_0, the isotropic scale of medium inhomogeneities:

$$v \sim V \sim \frac{1}{l_0}.$$
(2.76)

One of the coefficients v and V can be excluded by performing an appropriate unitary transform $\widehat{U}$ over the two-dimensional tensor $\widehat{\nu}$ and vector $\boldsymbol{\Gamma} = \{\Gamma_1, \Gamma_2\}$:

$$\begin{pmatrix} \Gamma'_1 \\ \Gamma'_2 \end{pmatrix} = \hat{U} \begin{pmatrix} \Gamma_1 \\ \Gamma_2 \end{pmatrix}, \quad \hat{\nu}' = \hat{U}\hat{\nu}\hat{U}^+. \tag{2.77}$$

In particular, by choice of parameters α and β in the transformation

$$\hat{U} = \begin{pmatrix} \cos\alpha & e^{i\beta}\sin\alpha \\ -e^{-i\beta}\sin\alpha & \cos\alpha \end{pmatrix}, \tag{2.78}$$

the parameter V may be set to zero. It is an easy matter to verify that a new expression for v' will then be

$$v' = (v^2 + |V|^2)^{1/2}. \tag{2.79}$$

After having accomplished the transformations noted in the QIA equations (2.48), one arrives at

$$\frac{d\Gamma_1}{ds} = \frac{1}{2}ik_0\varepsilon_0^{-1/2}(vs\Gamma_1 + \nu_{12}\Gamma_2),$$

$$\frac{d\Gamma_2}{ds} = \frac{1}{2}ik_0\varepsilon_0^{-1/2}(\nu_{21}\Gamma_1 - vs\Gamma_2). \tag{2.80}$$

Here we specially shifted the origin of s to give $\nu_{11} = 0$ at $s = 0$, and, for brevity, omitted the primes with all quantities $\Gamma'_{1,2}$, v', ν'_{12}, ν'_{21}, ν'_{21} and the index "0" coming with ν^0_{12}.

If the distance given from a "particular" point $s = 0$, where $\nu_{11} = 0$, is large, the component $\nu_{11} = vs$ considerably exceeds the component ν_{12}. For an initially extraordinary wave ($n_e = n_1 < n_2$) this makes it possible to formulate the initial condition in the form

$$|\Gamma_1(-\infty)| = |\Gamma^e_1(-\infty)| = 1, \quad |\Gamma_2(-\infty)| = |\Gamma^e_2(-\infty)| = 0. \tag{2.81}$$

In the case when the incident wave is an ordinary one ($n_o = n_2 > n_1$) the initial condition is specified in another way:

$$|\Gamma_1(-\infty)| = |\Gamma^o_1(-\infty)| = 0, |\Gamma_2(-\infty)| = |\Gamma^o_2(-\infty)| = 1. \tag{2.82}$$

An important advantage of Eqs. (2.80) is that they yield to analytical solution by the Landau–Zener method.

Introduce the dimensionless variable

$$\xi = \left(\frac{k_0 v}{2\varepsilon_0^{1/2}} \right)^{1/2} s \tag{2.83}$$

in Eqs. (2.80) and redefine Γ_2 as

$$\Gamma_2 \rightarrow \Gamma_2 \nu_{12}/|\nu_{12}|. \tag{2.84}$$

Then Eqs. (2.80) convert to the system

$$\frac{d\Gamma_1}{d\xi} = i\xi\Gamma_1 + i\frac{p^{1/2}}{2}\Gamma_2, \quad \frac{d\Gamma_2}{d\xi} = i\frac{p^{1/2}}{2}\Gamma_1 - i\xi\Gamma_2, \tag{2.85}$$

which contains only a single parameter

$$p = 2\varepsilon_0^{-1/2} k_0 v^{-1} |\nu_{12}|^2 = 2\varepsilon_0^{-1/2} k_0 \frac{|\nu_{12}|^2}{|\partial \nu_{11}/\partial s|}. \tag{2.86}$$

The physical sense of the parameter p is that it defines the phase delay $\Delta\varphi$ between normal waves, which takes place at the interval of linear interaction between these waves l_{int}. This may be assessed from the condition

$$|\nu_{11}| \sim |\nu_{12}| \quad or \quad vl_{\text{int}} \sim |\nu_{12}|, \tag{2.87}$$

whence

$$l_{\text{int}} \sim \frac{|\nu_{12}|}{v}. \tag{2.88}$$

The scale l_{int} is a particular case of the polarization scale of inhomogeneity, $l_p(\mathbf{r})$, introduced in Section 1.2. More precisely, l_{int} is the magnitude of $l_p(\mathbf{r})$ taken at the minimum of $|n_1 - n_2|$.

Supposing the interaction length l_{int} is known, we may estimate the phase delay between the normal waves over this interval, $\Delta\varphi_{\text{int}} = k_0\Delta n l_{\text{int}}$, by putting

$$\Delta n = \Delta n_{min} = \frac{|\nu_{12}|}{2n_0}.$$

As a result, we conclude that the quantity $\Delta\varphi$ is indeed comparable with p. It is this parameter which alone defines the coefficient of mutual transformation of normal waves due to the localized interaction.

Note that the interaction length l_{int} could be introduced in a somewhat different manner, through the refractive indices of the "simplified"

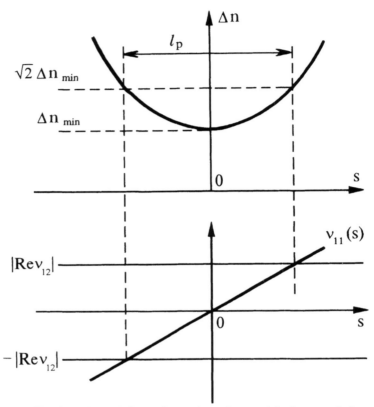

FIGURE 2.3. The length l_p defines the region of essential change of the polarization of electromagnetic field

normal waves. According to (2.69), at $\nu_{11} = -\nu_{22} = vs$ the difference between the refractive indices of two "simplified" waves is

$$\Delta n = \frac{1}{2n_0}[(vs)^2 + |\nu_{12}|^2]^{1/2}, \quad n_0 = \varepsilon_0^{1/2}.$$

Let $\Delta n_{min} = |\nu_{12}|/2n_0$ be a minimum magnitude of Δn in the inter-action region. The distance over which Δn exceeds Δn_{min} by a factor of $\sqrt{2}$ (see Figure 2.3) corresponds to vs and $|\nu_{12}|$ being equal; hence the interaction length l_{int} can be estimated from

$$\Delta n(l_{int}) \approx \sqrt{2}\Delta n_{min}.$$

Elimination of Γ_1 from (2.85) results in the Weber–Hermite equation

$$\frac{d^2\Gamma_2}{d^2\xi^2} + \left(\frac{p}{4} + i + \xi^2\right)\Gamma_2 = 0, \tag{2.89}$$

whose solution is expressed through the functions of the parabolic cylinder $D_n(z)$ (Whittaker and Watson, 1965). Such a solution has been used by Landau (1932) and Zener (1932) in a related problem on electron term interaction.

Eqs. (2.85) and condition (2.81) (we assume that an extraordinary wave approaches the region of polarization transformation) are satisfied by functions

$$\Gamma_1^e(\xi) = \Gamma_1(\xi) = e^{-\pi p/32 + i\pi/4} D_{ip/8}(\sqrt{2}\, e^{3\pi i/4}\xi), \tag{2.90}$$

$$\Gamma_2^e(\xi) = \Gamma_2(\xi) = \left(\frac{p}{8}\right)^{1/2} e^{-\pi p/32} D_{ip/8-1}(\sqrt{2}\, e^{3\pi i/4}\xi). \tag{2.91}$$

Taking into account asymptotic formulae for $D_n(z)$ at $z \to \infty$ (Whittaker and Watson, 1965), we find the intensities of extraordinary, $|\Gamma_1|^2$, and ordinary, $|\Gamma_2|^2$ waves at $\xi \to +\infty$:

$$|\Gamma_1(+\infty)|^2 = e^{-\pi p/4}, \quad |\Gamma_2(+\infty)|^2 = 1 - e^{-\pi p/4}. \tag{2.92}$$

Accordingly, the coefficient of mutual transformation of extraordinary waves into ordinary ones is

$$\zeta = \exp(-\pi p/4). \tag{2.93}$$

In the case that the incident wave is an ordinary one the solution should be sought subject to the initial condition (2.82). Then instead of (2.90)–(2.92) it follows that Γ_1 is given by the expression complex conjugate to (2.91), while Γ_2 is given by (2.90) taken with the opposite sign and complex conjugated. The coefficient of transformation of an ordinary wave into the extraordinary one in this case is also $\exp(-\pi p/4)$ which agrees with the reciprocity theorem.

The solution of QIA equations with linearized coefficients converts into non-interacting normal waves at $s \gg l_{\text{int}}$. Therefore, matching solutions (2.90) and (2.91) with normal waves corresponding to the initial system (2.48) provides a uniform asymptotic solution over the whole line of numbers. For the matching procedure to be applicable we need the approximation of independent normal waves to already hold at the end of the region of coefficient linearization.

2.5.5   ASYMPTOTIC APPROACHES TO SOLUTION OF QIA EQUATIONS

As noted above, the system of coupled equations (2.71) admits exact solutions in the case of constant coefficients and coefficients varying linearly with time. Each of them can be used as a basis for constructing asymptotic solutions.

An asymptotic solution to (2.71) exploiting the slowness of variation of the coefficients a_{ij} assumes the form of harmonic oscillations with slowly varying amplitudes and phases. Such a solution emerges provided that the coefficients b_{ij} alter little for the typical period of oscillations $T_\Omega = 2\pi/\Omega$, i.e. when

$$T_\Omega \left| \frac{db_{ij}}{dt} \right| \ll 1. \tag{2.94}$$

For QIA equations (2.48), (2.94) corresponds to

$$\frac{1}{\delta} = \frac{\mu}{\mu_1} \ll 1.$$

Indeed, the requirement for components of the anisotropy tensor ν_{ij} to vary only by a small amount over the period $l_b = 1/k_0\Delta n$ of spatial beats between modes is expressed by the inequality

$$l_b \left| \frac{\partial \nu_{ij}}{\partial s} \right| / \max |\nu_{ij}| \sim \frac{1}{k_0\mu_1} \cdot \frac{\mu_1}{l} \cdot \frac{1}{\mu_1} \sim \frac{\mu}{\mu_1} = \frac{1}{\delta} \ll 1.$$

Here Δn is the difference in the refractive indices of normal modes, which, according to (2.69), equals

$$\Delta n = |n_1 - n_2| = \frac{1}{2}\varepsilon_0^{-1/2} \left[\frac{(\nu_{11} - \nu_{22})^2}{4} + |\nu_{12}|^2 \right]^{1/2},$$

and scales as $\mu_1 = \max |\nu_{mn}|$.

If $\delta \gg 1$ the QIA equations admit a solution in the form of series in inverse powers of δ, see Wazow (1965), with the leading term of the asymptotic series corresponding to the independent "simplified" normal waves described above in Section 2.4.6. We will exploit asymptotic solutions of that kind as a germ in the split ray method (Section 3.3) and the synthetic method (Section 3.4).

For $\delta \leq 1$ the interaction between normal waves becomes significant. To describe it one may conveniently use the QIA equations in the form of the coupled equations for amplitudes of interacting modes. Such equations are derived and analyzed below, in Section 3.2.

The Landau–Zener solution described in the previous section may also be used as a basis for an asymptotic solution. Instead of exponential functions in the case $a_{ij} = $ const, in this case the role of standard solution is performed by the functions of a parabolic cylinder (2.90) and (2.91) which satisfy (2.71) with linearly varying coefficients. Provided the coefficients a_{ij} deviate smoothly from the linear law the solution will preserve the behavior characteristic of the functions of a parabolic cylinder. One can only expect that arguments of those functions together with amplitude coefficients will undergo specific changes, as happens with a solution of a related problem involving wave transmission over a barrier, whose form qualitatively resembles a quadratic dependence (Olver, 1954, 1958). This procedure, although inherently clear, has still not been implemented adequately.

3

Modifications and generalizations of QIA

3.1 QIA equations for the vector of electric induction

The assumption that the field $\mathcal{E}$ is transverse adopted in the primary version of the QIA holds true only in the framework of the zeroth approximation. In reality, the transverse field component E_3 differs from zero and, in principle, could be found in the first approximation by the parameter μ_1. Nevertheless, if one formulates the QIA equations for the induction vector $\mathcal{D}$, the transverse component E_3 can already be found in the zeroth approximation. The relevant modification of the QIA was implemented by Naida (1977a, 1978b).

In the framework of the standard QIA the relation between the vectors **D** and **E** in the zeroth approximation is

$$D_1 = \varepsilon_{11} E_1 + \varepsilon_{12} E_2, \qquad D_2 = \varepsilon_{21} E_1 + \varepsilon_{22} E_2, \qquad (3.1)$$

whereas in reality they are connected by the relationships

$$D_1 = \varepsilon_{11} E_1 + \varepsilon_{12} E_2 + \varepsilon_{13} E_3 = \det(\hat{\varepsilon}) \frac{\chi_{22} E_1 - \chi_{12} E_2}{\varepsilon_{33}},$$

$$D_2 = \varepsilon_{21} E_1 + \varepsilon_{22} E_2 + \varepsilon_{23} E_3 = \det(\hat{\varepsilon}) \frac{-\chi_{21} E_1 + \chi_{11} E_2}{\varepsilon_{33}}, \qquad (3.2)$$

which follow from the transversality of the vector **D** with respect to the wave-vector **k**, i.e. the condition $D_3 = 0$. For the most part, the transition from (3.2) to (3.1) does not introduce serious errors. Nevertheless, there exist some important cases when terms quadratic in μ_1, which are omitted in the transition from exact expressions (3.2) to their approximate counterparts (3.1), are comparable to terms linear in μ_1 which are left in (3.1). In just such cases it is expedient to deal with exact equations (3.2).

Without running the risk of impairing the accuracy of the QIA, we replace the multiplier $\det(\hat{\varepsilon})/\varepsilon_{33}$ in (3.2) by ε_0^2 and take this into account when passing from Eqs. (2.30) to the QIA equations for the vector $\mathbf{D}$. As a result we obtain the following expression for the electric induction vector

$$D = \mathbf{D} \exp\left(ik_0 \int n_0 ds\right), \quad \mathbf{D} = \Phi_0 \varepsilon_0^{3/4}(\Gamma_1 \mathbf{q}_1 + \Gamma_2 \mathbf{q}_2),$$

which is analogous to Eqs. (2.61). The differences are in the additional factor ε_0 coming with the amplitude Φ_0 (since $\mathbf{D} \cong \varepsilon_0 \mathbf{E}$), and in the fact that the normalized components $\Gamma_{1,2} = D_{1,2}/(\Phi_0 \varepsilon_0^{3/4})$ obey the coupled QIA equations (2.62) except for a replacement of the anisotropy tensor components $\nu_{ik} = \varepsilon_{ik} - \varepsilon_0 \delta_{ik}$ entering them by components of the modified anisotropy tensor

$$\tilde{\nu}_{ik} = \varepsilon_0^2(\varepsilon_0^{-1}\delta_{ik} - \chi_{ik}). \tag{3.3}$$

The latter characterizes the degree of anisotropy of the inverse tensor of the dielectric permittivity $\hat{\chi} = \hat{\varepsilon}^{-1}$. Also, we will need the following components of the tensor $\tilde{\nu}_{ik}$:

$$\tilde{\nu}_{11} = \varepsilon_0^2(\varepsilon_0^{-1} - \chi_{11}), \qquad \tilde{\nu}_{12} = -\varepsilon_0^2 \chi_{12},$$

$$\tilde{\nu}_{21} = -\varepsilon_0^2 \chi_{21}, \qquad \tilde{\nu}_{22} = \varepsilon_0^2(\varepsilon_0^{-1} - \chi_{22}).$$

The modified tensor $\tilde{\nu}_{ik}$ departs from the original tensor ν_{ik} only in the second order in the small parameter μ_1, i.e

$$\tilde{\nu}_{ik} = \nu_{ik} + O(\mu_1^2),$$

since

$$\chi_{ik} = (\hat{\varepsilon}^{-1})_{ik} = \varepsilon_0^{-1}\delta_{ik} - \varepsilon_0^{-2}(\varepsilon_{ik} - \varepsilon_0 \delta_{ik}) + \mathcal{O}(\mu_1^2)$$

$$= \varepsilon_0^{-1}\delta_{ik} - \varepsilon_0^{-2}\nu_{ik} + \mathcal{O}(\mu_1^2).$$

As before, for unit vectors $\mathbf{q}_1$ and $\mathbf{q}_2$ one may safely use arbitrary orthonormalized vectors forming with $\mathbf{k}$ a right triple $\mathbf{q}_1, \mathbf{q}_2, \mathbf{k}$. From now on it will be assumed that $\tilde{\mathbf{q}}_1$ and $\tilde{\mathbf{q}}_1$ are the unit eigenvectors of the two-dimensional tensor

$$\mathrm{Re}\begin{pmatrix} \chi_{nn} & \chi_{nb} \\ \chi_{bn} & \chi_{bb} \end{pmatrix}. \tag{3.4}$$

As in the case of the QIA equations for the $\mathcal{E}$ vector, the electric induction vector $\mathcal{D}$ may be represented in an invariant form

$$\mathcal{D} = \tilde{\mathbf{D}} \exp\left(\frac{1}{4}ik_0 \int \varepsilon_0^{3/2}(6\varepsilon_0^{-1} - \chi_{11} - \chi_{22})\, ds\right),$$

(3.5)

$$\tilde{\mathbf{D}} = \Phi_0 \varepsilon_0^{3/4}(\tilde{\Gamma}_1 \mathbf{q}_1 + \tilde{\Gamma}_2 \mathbf{q}_2),$$

which is the least sensitive to the choice of the ε_0 value. Here Φ_0 as usual satisfies the energy conservation law (2.37) whereas $\tilde{\Gamma}_1$ and $\tilde{\Gamma}_2$ obey Eqs. (2.66) in which the coefficients γ_{ik} defined by (2.66a) should be replaced with

$$\tilde{\gamma}_{11} = -\tilde{\gamma}_{22} = \varepsilon_0^2(\chi_{22} - \chi_{11})/2, \qquad \tilde{\gamma}_{12} = -\varepsilon_0^2 \chi_{12},$$

(3.6)

$$\tilde{\gamma}_{21} = -\varepsilon_0^2 \chi_{21}.$$

Ultimately, the invariant equations (2.66) have the following form (Naida, 1977a)

$$\frac{d\tilde{\Gamma}_1}{ds} + \frac{i}{2}k_0\varepsilon_0^{3/2}\left(\frac{\chi_{11} - \chi_{22}}{2}\tilde{\Gamma}_1 + \chi_{12}\tilde{\Gamma}_2\right) - T_{\text{eff}}^{-1}\tilde{\Gamma}_2 = 0,$$

(3.7)

$$\frac{d\tilde{\Gamma}_2}{ds} + \frac{i}{2}k_0\varepsilon_0^{3/2}\left(\chi_{21}\tilde{\Gamma}_1 + \frac{\chi_{22} - \chi_{11}}{2}\tilde{\Gamma}_2\right) + T_{\text{eff}}^{-1}\tilde{\Gamma}_1 = 0.$$

The parameter T_{eff}^{-1} stands for the effective ray torsion and is expressed by (2.63).

On determining the components $\tilde{\Gamma}_{1,2}$ from the equations written above one readily obtains the electric field strength $\mathcal{E} = \hat{\chi}\mathcal{D}$:

$$\mathcal{E} = \Phi_0 \varepsilon_0^{3/4} \hat{\chi}\left[\Gamma_1 \mathbf{q}_1 + \Gamma_2 \mathbf{q}_2 \exp\left(ik_0 \int n_0\, ds\right)\right],$$

(3.8)

or, in invariant form,

$$\mathcal{E} = \Phi_0 \varepsilon_0^{3/4} \hat{\chi}\left(\tilde{\Gamma}_1 \mathbf{q}_1 + \tilde{\Gamma}_2 \mathbf{q}_2\right)$$

(3.9)

$$\times \exp\left(\frac{ik_0}{4} \int \varepsilon_0^{3/2}(6\varepsilon_0^{-1} - n_1^{-2} - n_2^{-2})\, ds\right).$$

By exploiting (2.7), we replaced the sum $\chi_{11} + \chi_{22}$ here by $n_1^{-2} + n_2^{-2}$.

Abandoning the standard QIA equations for the electric field strength vector in favor of formulae (3.3)–(3.9) for the induction vector allows us to expand the QIA's region of applicability. Specifically, this occurs if the condition (2.70) is rejected. These formulae guarantee a continuous transition to a homogeneous anisotropic medium, albeit in the form of "simplified" normal waves, and make it possible to write equations for the case of a 3D inhomogeneous medium which are almost as general as Budden's equations in the case of a plane-layered medium. Also, they describe ray splitting into extraordinary and ordinary rays, and, finally, appropriately match the Courant–Lax formulae in those regions where the anisotropy is weak ($\mu_1 \ll 1$), but condition (2.23), $\delta = \mu_1/\mu \gg 1$, holds. The scheme outlined will be developed in Sections 3.2 and 3.3.

3.2   QIA equations in the form of equations for interacting modes

3.2.1   TRANSFORMED EQUATIONS FOR AMPLITUDES OF NORMAL MODES

Let us rewrite the QIA equations in a form describing the interaction of normal waves. With this in mind, we take real unit eigenvectors of the two-dimensional tensor (3.4) for real unit vectors $\mathbf{q}_1$ and $\mathbf{q}_2$ and pass to the unit eigenvectors $\mathbf{e}_{1\perp}$ and $\mathbf{e}_{2\perp}$ of the two-dimensional tensor $\chi_{\alpha\beta}$ which play the role of polarization vectors.

Passing to the normal waves can be accomplished in two different, yet analogous ways, depending on which of the waves, ordinary or extraordinary, we want to approach. In both cases we shall depart from formulae (3.8) and (3.9), while modifying the eikonal to have the form $k_0 \int n_1\, ds$ for the extraordinary wave and $k_0 \int n_2\, ds$ for the ordinary one (up to terms of the μ_1^2 order in the integrand). Then the total field $\mathcal{E}$ can be (now only formally) written in the form of the extraordinary wave

$$\mathcal{E} = \Phi_0 n_0^{3/2} (n_1^{-2} C_1 \mathbf{e}_1 + n_2^{-2} C_2 \mathbf{e}_2)$$

$$\times \exp\left[ik_0 \int \left(\frac{3}{2} n_0 - \frac{1}{2} n_0^3 n_1^{-2}\right) ds\right]. \tag{3.10}$$

Here, as before, ds implies the element of the length of a non-split ray corresponding to the Hamiltonian system of equations (2.12) with $n = n_0(\mathbf{r})$. The refractive indices n_1 and n_2 and the polarization unit vectors $\mathbf{e}_1$ and $\mathbf{e}_2$ correspond, by virtue of Eqs. (2.18)–(2.20), to the current value of the dielectric permittivity tensor $\hat{\varepsilon}(\mathbf{r})$ and to the magnitude of the wave-vector $\mathbf{k} = n_0 k_0 \boldsymbol{\tau}$, where $\boldsymbol{\tau}$ is the tangent unit vector to the non-split ray. The coefficients C_1 and C_2 appearing in Eq. (3.10) are linked to $\tilde{\Gamma}_1$ and $\tilde{\Gamma}_2$ by the transform

$$\tilde{\Gamma}_1 = (e_{11}C_1 + e_{21}C_2) \exp \left[\frac{1}{4} i k_0 \int n_0^3 (n_2^{-2} - n_1^{-2}) \, ds \right] ,$$

$$\tilde{\Gamma}_2 = (e_{12}C_1 + e_{22}C_2) \exp \left[\frac{1}{4} i k_0 \int n_0^3 (n_2^{-2} - n_1^{-2}) \, ds \right] ,$$

(3.11)

where the coefficients e_{ij} are defined by Eqs. (2.19).

After the aforementioned transformations Eqs. (3.7) reduce to the equations for the determination of the amplitude coefficients C_1 and C_2 (Naida, 1977a):

$$\frac{dC_1}{ds} + p_{11}C_1 + p_{12}C_2 = 0 ,$$

(3.12)

$$\frac{dC_2}{ds} + \frac{1}{2} i k_0 n_0^3 (n_2^{-2} - n_1^{-2}) C_2 + p_{21}C_1 + p_{22}C_2 = 0 .$$

Here we denote

$$p_{11} = p_{22}^* = \frac{2iJK}{1 + K^2} T_{\text{eff}}^{-1} = -\frac{iJ}{(1 + Q^2)^{1/2}} T_{\text{eff}}^{-1} ,$$

$$p_{12} = -p_{21}^* = i\Psi + J \frac{K^2 - 1}{1 + K^2} T_{\text{eff}}^{-1} = i\Psi - \frac{JQ}{(1 + Q^2)^{1/2}} T_{\text{eff}}^{-1} .$$

(3.13)

The values J, K, and Q are given by formulae (2.20), but the role of the vector $\mathbf{t}$ in these formulae is to be played by the unit vector $\boldsymbol{\tau}$ tangent to the "quasi-isotropic" ray. Accordingly, here

$$\Psi = -\frac{1}{2} \frac{d}{ds} \arctan Q .$$

The coupled equations (3.12) for amplitudes C_1 and C_2 describe the interaction of normal modes 1 and 2. In a homogeneous medium where $p_{\alpha\beta} = 0$, Eqs. (3.13) admit the solution $C_1 = \text{const} \neq 0$, $C_2 = 0$ while the field (3.10) conforms to the normal (extraordinary) wave $\mathcal{E}_1$ described by formulae (2.66), (2.67).

Similarly, the total field can also be expressed formally as the ordinary wave

$$\mathcal{E} = \Phi_0 n_0^{3/2} (n_1^{-2} C_1' e_1' + n_2^{-2} C_2' e_2')$$

$$\times \exp\left[ik_0 \int \left(\frac{3}{2} n_0 - \frac{1}{2} n_0^3 n_2^{-2} \right) ds \right].$$

(3.14)

This expression follows from Eq. (3.9) subject to the transform

$$\tilde{\Gamma}_1 = (e_{11} C_1' + e_{21} C_2') \exp\left[\frac{1}{4} ik_0 \int n_0^3 (n_2^{-2} - n_1^{-2}) \, ds \right],$$

$$\tilde{\Gamma}_2 = (e_{12} C_1 6 \prime + e_{22} C_2') \exp\left[\frac{1}{4} ik_0 \int n_0^3 (n_2^{-2} - n_1^{-2}) \, ds \right],$$

(3.15)

As a result of this transform Eqs. (3.7) reduce to the equations for the amplitude coefficients C_1' and C_2':

$$\frac{dC_1'}{ds} + \frac{1}{2} ik_0 n_0^3 (n_1^{-2} - n_2^{-2}) C_1' + p_{11} C_1' + p_{12} C_2' = 0,$$

$$\frac{dC_2'}{ds} + p_{21} C_1' + p_{22} C_2' = 0,$$

(3.16)

with coefficients p_{11}, p_{12}, p_{21}, p_{22} being given by formula (3.13) as previously.

With the accuracy given up to the terms of the μ_1^2 order, the arguments of exponents in Eqs. (3.10) and (3.14) are equal, respectively, to

$$ik_0 \int n_1 \, ds \quad \text{and} \quad ik_0 \int n_2 \, ds.$$

Notice that the arguments of exponents in Eqs. (3.10) and (3.11) become exact on choosing $n_0 = n_1$ whereas in Eqs. (3.14) and (3.15) this occurs for $n_0 = n_2$.

It can readily be shown that the wave field (3.14) with coefficients C_1' and C_2' calculated from (3.16) is equivalent to the field (3.10) with amplitudes calculated from (3.12). The distinction between Eqs. (3.12) and (3.16) is that Eqs. (3.12) are preferable in calculations of extraordinary waves ($\phi = ik_0 \int n_1 \, ds$) whereas (3.16) fit the case of ordinary waves ($\phi = ik_0 \int n_2 \, ds$).

3.2.2 DEFORMED NORMAL WAVES

In the region of comparatively strong birefringence $\delta \gg 1$ (weak interaction of normal modes), one can construct solutions to (3.12) and (3.16) that very closely correspond to the normal waves (2.15). For that purpose we invoke the iterative approach proposed by Naida (1977a). In the zeroth approximation we assume that there exists a single normal wave in original form (2.15). That wave we shall take as a "seed" for the iterative procedure. Say, the extraordinary wave (with index 1) will serve as a seed at $C_1^{(0)} \neq 0$ and $C_2^{(0)} = 0$, whereas the ordinary wave will do so at $C_1^{(0)} = 0$ and $C_2^{(0)} \neq 0$. In the region of weak wave interaction, i.e., at $\delta = \mu_1/\mu \sim |n_1 - n_2| k_0 l \gg 1$, the coefficient $k_0(n_1 - n_2)$ is large compared to $1/l$: $|n_1 - n_2| k_0 \gg 1/l$. Hence, a formal asymptotic expansion in inverse powers of the large parameter $\delta = k_0(n_1 - n_2)l \gg 1$ can be devised for C_1 and C_2. Here l is the scale of variation of the parameters of the medium.

In the case of the "extraordinary" seed this implies that initial terms in the expansion of C_α in a series in inverse powers of large parameter $\delta \gg 1$,

$$C_\alpha = C_\alpha^{(0)} + C_\alpha^{(1)} + C_\alpha^{(2)} + \ldots \quad (\alpha = 1, 2) \qquad (3.17)$$

should be subject to the condition

$$C_1^{(0)}(s_{\text{in}}) \neq 0, \qquad C_2^{(0)} \equiv 0. \qquad (3.18)$$

Since the coefficient $C_2^{(0)}$ is equal to zero, the first of the equations (3.12) is simplified and defines the behavior of the coefficient $C_1^{(0)}$:

$$\frac{dC_1^{(0)}}{ds} + p_{11}C_1^{(0)} = 0.$$

Subsequent terms of the expansion (3.17) are determined from the recurrence formulae

$$C_2^{(m)} = 2ik_0^{-1}n_0^{-3}(n_2^{-2} - n_1^{-2})^{-1}\left[\frac{dC_2^{(m-1)}}{ds} + p_{21}C_1^{(m-1)} + p_{22}C_2^{(m-1)}\right],$$
$$\qquad (3.19)$$

$$\frac{dC_1^{(m)}}{ds} + p_{11}C_1^{(m)} + p_{12}C_2^{(m)} = 0,$$

$$C_1^{(m)}(s_{\text{in}}) \equiv 0 \quad (m \geq 1).$$

A similar procedure can also be applied in respect of equations with the "ordinary" seed $C_2^{(0)} \neq 0$ and $C_1^{(0)} \equiv 0$.

The coefficients that one finds for the "extraordinary" seed (3.18) in the region of relatively strong birefringence $\delta = k_0 l |n_1 - n_2| \gg 1$ satisfy the estimate

$$\frac{|C_2|}{|C_1|} \lesssim \frac{1}{\delta} = |n_1 - n_2|^{-1} k_0^{-1} l^{-1} = \frac{\mu}{\mu_1}. \tag{3.20}$$

If this condition is observed, the replacement of n_2^{-2} by n_1^{-2} in the term with small amplitude C_2 in (3.10) will not involve large errors. When this is done, (3.10) takes the form

$$\mathcal{E} = \Phi_0 n_0^{3/2} n_1^{-2} (C_1 e_1 + C_2 e_2) \exp\left(i k_0 \int (\frac{3}{2} n_0 - \frac{1}{2} n_0^3 n_1^{-2}) \, ds \right). \tag{3.21}$$

The distinction between this expression and (3.12) is estimated as

$$\max \frac{|\Delta C_{1,2}|}{|C_1|} \lesssim \mu = \frac{1}{k_0 l} \ll 1. \tag{3.22}$$

Within the same insignificant error (3.22) one may use the substitution

$$n_0^3 (n_2^{-2} - n_1^{-2}) \to 2(n_1 - n_2)$$

in Eqs. (3.12), (3.16) and (3.19).

For a relatively strong birefringence, $\delta \gg 1$, the amplitude component C_2 in the solution (3.21) obtained with the seed (3.17)–(3.19) appears as a small correction. If it can be omitted, (3.21) has the structure of the original normal wave (2.15). If we set $n_0 = n_1$ in (3.21), this will not affect the accuracy. As a result, the QIA formula (3.21) is simplified and takes the form

$$\mathcal{E}_e = \Phi_0 n_1^{-1/2} (C_1 e_1 + C_2 e_2) \exp\left(i k_0 \int n_1 \, ds \right). \tag{3.23}$$

Similar operations with the ordinary seed $C_1^{(0)} = 0$, $C_2^{(0)} \neq 0$ lead to

$$\mathcal{E}_o = \Phi_0 n_2^{-1/2} (C_1' e_1 + C_2' e_2) \exp\left(i k_0 \int n_2 \, ds \right). \tag{3.24}$$

It is of importance that the amplitude factors of both solutions (3.23) and (3.24) do not contain oscillating terms for $\delta = \mu_1/\mu \gg 1$. Oscillations in the amplitudes of waves (3.23) and (3.24) also do not occur in the region of relatively weak birefringence, $\delta = \mu_1/\mu \lesssim 1$, where the iterations cease to converge. This is for another reason, which is the smallness of the parameter $\delta \lesssim 1$. Hence wave solutions (3.23) and (3.24), by their phase structure, are very close to normal waves (2.15) either for $\delta \gg 1$ or $\delta \lesssim 1$, but only up to their second encounter with the region of a sufficiently strong birefringence.

In other words, in the above-mentioned regions waves (3.23) and (3.24) possess the phase structure of normal waves (2.15), but have a deformed polarization structure compared to that of (2.15). The deformation of polarization is relatively weak for $\delta \gg 1$, but grows (up to 100%) for $\delta \lesssim 1$. We call the solutions (3.23) and (3.24) *deformed normal waves*.

Therefore, the coefficient C_2 in Eq. (3.23) loses the sense of the amplitude of the ordinary wave for deformed normal waves, and only defines a correction to the polarization of the extraordinary wave. However if the conditions (3.17)–(3.19) are not satisfied, C_2 again assumes the sense of the amplitude of the ordinary wave and oscillates as $\exp[ik_0 \int (n_2 - n_1)\, ds]$. The coefficient C_1' in Eq. (3.24) possesses analogous properties.

Using deformed normal waves, we can readily establish a connection between the QIA and the Budden equations and describe the process of splitting waves into ordinary and extraordinary waves.

3.2.3   EQUATIONS FOR INTERACTING MODES. THEIR RELATION TO THE BUDDEN EQUATIONS

If small corrections of the order of μ are completely ignored, formula (3.21) for the extraordinary seed and the respective formula for the ordinary seed can be presented in a unified way:

$$\mathcal{E} = \Phi_0(n_1^{-1/2} F_1 \mathbf{e}_1 + n_2^{-1/2} F_2 \mathbf{e}_2)\,. \tag{3.25}$$

Here the amplitudes

$$F_1 = C_1 \exp\left(ik_0 \int n_1\, ds\right) + C_1' \exp\left(ik_0 \int n_2\, ds\right),$$

$$F_2 = C_2 \exp\left(ik_0 \int n_1\, ds\right) + C_2' \exp\left(ik_0 \int n_2\, ds\right)$$

satisfy the QIA equations in the form of *interacting normal waves*

$$\frac{dF_1}{ds} - ik_0 n_1 F_1 + p_{11} F_1 + p_{12} F_2 = 0 \,,$$

$$(3.26)$$

$$\frac{dF_2}{ds} - ik_0 n_2 F_2 + p_{21} F_1 + p_{22} F_2 = 0 \,.$$

For $\mu_1 \ll 1$, in a plane-layered medium these equations reduce to Budden's equations for interacting waves (1961).

Thus, the QIA matches quite naturally the classical Budden and Courant–Lax methods. The virtues of the QIA are that it applies not only to a plane-layered medium, as the Budden method does, but also to 3D inhomogeneous media. The drawback is that the QIA equations are limited by the condition of weak anisotropy, $\mu_1 \ll 1$, while the Budden and Courant–Lax methods are free of these limitations. In Section 3.4 we show how to modify the QIA equations in order to also fit the case of strong anisotropy $\mu_1 \sim 1$.

3.3   The method of split rays

3.3.1   BASIC PRINCIPLES

With all the superficial similarity between the QIA expressions (3.23) and (3.24) and independent normal waves of the Courant–Lax method, a marked distinction exists between them: the QIA ignores splitting of the rays into ordinary and extraordinary ones. Accordingly, their phases depart from those given by the exact expressions (2.13). However both these shortcomings can be eliminated relatively easily. The modification of the QIA, suggested by Naida (1977a, 1978b) serves that purpose. We call it *the method of split rays*. The basic idea behind that approach is to abandon the isotropic rays (dashed line in Figure 1.1) and replace them by the split rays corresponding to normal waves (Figure 1.1). On these rays, deformed normal waves should further be constructed with the help of equations of the type of (3.7), (3.12) and (3.16). The ray splitting occurs every time as a natural consequence of matching a single incident deformed normal wave with two waves leaving the interaction region $\delta \lesssim 1$. Thus the condition $\delta \lesssim 1$ bounds the region where wave interaction occurs and, simultaneously, ray splitting takes place. Using the concept of Fresnel zones enclosing the rays and defining their physical "thickness" (Kravtsov, Orlov 1980, 1990; Kravtsov 1988) it is possible to see that it is impossible to discriminate experimentally between the interacting normal waves in the region $\delta \lesssim 1$, or, equivalently, to discern the split rays. In this region none of the physical experiments (with slits, mirror antennae, travelling wave antennae, multipole receivers) is capable of distinguishing the split rays. They have a distinct existence of their own, which can be detected with physical instruments, only in the region of strong birefringence $\delta \gg 1$.

3.3.2   A FORMAL PROCEDURE OF RAY SPLITTING

In order to obtain the geometric-optical equations for the field on split rays one has to return again to the procedure connected with Eqs. (2.30).

Any modification of the quasi-isotropic approximation which attempts to ensure the exact transition to non-interacting normal waves, should necessarily be based on the eikonal substitution (2.2) with the eikonals $\varphi = \varphi_1$ or $\varphi = \varphi_2$ corresponding to extraordinary or ordinary waves. To satisfy this requirement, we should choose in Eqs. (2.30) the refractive index n_1 for the extraordinary wave instead of n_0 (and n_2 for the ordinary wave) and, correspondingly, unit vectors $\mathbf{q}_1$, $\mathbf{q}_2$, and $\mathbf{t} = \mathbf{k}_1/|\mathbf{k}_1|$ for the extraordinary and $\mathbf{q}_1'$ $\mathbf{q}_2'$, and $\mathbf{t}' = \mathbf{k}_2/|\mathbf{k}_2|$ for the ordinary wave. In order to make use of the smallness of the anisotropy of the medium, it is convenient to select among the unit vectors $\mathbf{q}_1$ and $\mathbf{q}_2$ such vectors $\tilde{\mathbf{n}}$ and $\tilde{\mathbf{b}}$ which would be close, respectively, to the normal $\mathbf{n}$ and binormal $\mathbf{b}$ and satisfy the orthogonality condition $\tilde{\mathbf{n}} \perp \tilde{\mathbf{b}} \perp \mathbf{t}$. To be specific, a real unit vector that is perpendicular to both the wave vector $\mathbf{k}_1$ and the normal $\mathbf{n}$ to the ray with index "1" will be suitable for $\tilde{\mathbf{b}}$. The vector $\tilde{\mathbf{n}}$ is then introduced as a real unit vector perpendicular to $\mathbf{k}_1$ and $\tilde{\mathbf{b}}$ and added to them to form a right triple $\mathbf{k}_1$, $\tilde{\mathbf{n}}$, and $\tilde{\mathbf{b}}$. Similarly one may introduce vectors $\tilde{\mathbf{b}}'$ and $\tilde{\mathbf{n}}'$ corresponding to the wave vector $\mathbf{k}_2$, the normal $\mathbf{n}'$ and the binormal $\mathbf{b}'$ to the ray of index "2".

Apparently, in a weakly anisotropic medium the unit vectors $\mathbf{t} = \mathbf{k}_1/|\mathbf{k}_1|$, $\tilde{\mathbf{n}}$ and $\tilde{\mathbf{b}}$ are only different from the tangent to the ray $\boldsymbol{\tau}$, normal $\mathbf{n}$ and binormal $\mathbf{b}$ in small terms of order μ_1:

$$|\mathbf{t} - \boldsymbol{\tau}| \sim |\tilde{\mathbf{n}} - \mathbf{n}| \sim |\tilde{\mathbf{b}} - \mathbf{b}| \lesssim \mu_1 . \tag{3.27}$$

Similar estimations also hold for the ordinary ray.

We shall construct real moving unit vectors $\mathbf{q}_1$ and $\mathbf{q}_2$ of general form from $\tilde{\mathbf{n}}$ and $\tilde{\mathbf{n}}$ by applying formulae similar to (2.51):

$$\mathbf{q}_1 = \tilde{\mathbf{n}} \cos \psi + \tilde{\mathbf{b}} \sin \psi , \qquad \mathbf{q}_2 = -\tilde{\mathbf{n}} \sin \psi + \tilde{\mathbf{b}} \cos \psi .$$

For example, one may use the unit eigenvectors of two-dimensional real tensors

$$\text{Re} \begin{pmatrix} (\tilde{\mathbf{n}}, \widehat{\chi}\tilde{\mathbf{n}}) & (\tilde{\mathbf{n}}, \widehat{\chi}\tilde{\mathbf{b}}) \\ (\tilde{\mathbf{b}}, \widehat{\chi}\tilde{\mathbf{n}}) & (\tilde{\mathbf{b}}, \widehat{\chi}\tilde{\mathbf{b}}) \end{pmatrix},$$

$$\text{Re} \begin{pmatrix} (\tilde{\mathbf{n}}', \widehat{\chi}\tilde{\mathbf{n}}') & (\tilde{\mathbf{n}}', \widehat{\chi}\tilde{\mathbf{b}}') \\ (\tilde{\mathbf{b}}', \widehat{\chi}\tilde{\mathbf{n}}') & (\tilde{\mathbf{b}}', \widehat{\chi}\tilde{\mathbf{b}}') \end{pmatrix}$$

as unit vectors $\mathbf{q}_1$ and $\mathbf{q}_2$, and $\mathbf{q}_1'$, $\mathbf{q}_2'$.

By analogy with (3.9), we shall seek the solutions to (2.1) along extraordinary rays (with index 1) in the form

$$\mathcal{E} = \Phi_1 n_1^{3/2} \widehat{\chi}(\tilde{\Gamma}_1 \mathbf{q}_1 + \tilde{\Gamma}_2 \mathbf{q}_2) \exp\left[i\varphi_1 + \frac{1}{4} i k_0 \int n_1^3 (n_1^{-2} - n_2^{-2}) \, ds_1\right].$$

$$(3.28)$$

Here ds_1 is the element of the extraordinary ray length, φ_1 is the corresponding eikonal calculated by (2.13). The values of the refractive indices n_1 and n_2 in (3.28) are those on the extraordinary ray. They are functions of two vector arguments $\mathbf{r}$ and $\mathbf{t} = \mathbf{k}_1/|\mathbf{k}_1|$.

The amplitude factor Φ_1 satisfies the conservation law

$$\text{div}\,(\Phi_1^2 \mathbf{t}) = 0. \qquad (3.29)$$

This can be derived, say, by the well-known formula (Kravtsov and Orlov, 1990):

$$\Phi_1 = \left[\frac{D(x, y, z)}{D(s_1, \alpha_1, \beta_1)}\right]^{-1/2}, \qquad (3.30)$$

where x, y, and z are the Cartesian coordinates, α_1, β_1 are the parameters of the given family of extraordinary rays, for example, ray coordinates in the plane of origin. The dependence $\mathbf{r} = \mathbf{r}(s_1, \alpha_1, \beta_1)$ follows from the Hamiltonian system of equations (2.11). Of course, (3.30) also applies to an isotropic medium.

Now substitute (3.28) into (2.1) and project the resultant equation on $\tilde{\mathbf{n}}$ and $\tilde{\mathbf{b}}$, i. e., go to Eqs. (2.30). After that, we shall find the system of QIA equations for coefficients $\tilde{\Gamma}_1$ and $\tilde{\Gamma}_2$, which resembles (3.7), by making use of Rytov's identities (2.33) and neglecting the contributions of order μ_1^2 (Naida, 1977a, 1978b) :

$$\frac{d\tilde{\Gamma}_1}{ds_1} + \frac{i}{2}k_0 n_1^3 \left(\frac{\chi_{11} - \chi_{22}}{2}\tilde{\Gamma}_1 + \chi_{12}\tilde{\Gamma}_2\right) - T_{\text{eff}}^{-1}\tilde{\Gamma}_2 = 0 \,,$$

(3.31)

$$\frac{d\tilde{\Gamma}_2}{ds_1} + \frac{i}{2}k_0 n_1^3 \left(\chi_{21}\tilde{\Gamma}_1 + \frac{\chi_{22} - \chi_{11}}{2}\tilde{\Gamma}_2\right) + T_{\text{eff}}^{-1}\tilde{\Gamma}_1 = 0 \,.$$

The respective equations associated with the ordinary ray follow from (3.31) after the substitution of indices associated with ds and n.

Therefore, the QIA method is divided here into two branches that correspond to two normal waves (ordinary and extraordinary) and rely on the respective rays. It is easy to verify that in the region of relatively strong birefringence, $\delta \gg 1$, the method of split rays provides an asymptotic transition, at least with a correct phase, to non-interacting normal waves (2.15). The transition from the "split" version of the QIA to the non-interacting normal waves is accomplished by going through the same intermediate stages as when considering the non-split form.

First of all, one should pass from the polarization components of the field, $\tilde{\Gamma}_1$ and $\tilde{\Gamma}_2$, in Eqs. (3.28) and (3.31) (for extraordinary waves) to the polarization components C_1 and C_2 of deformed normal waves. This conversion is accomplished by formulae analogous to Eqs. (3.15):

$$\tilde{\Gamma}_1 = (e_{11}C_1 + e_{21}C_2)\exp\left[\frac{1}{4}ik_0\int n_1^3(n_2^{-2} - n_1^{-2})\,ds_1\right]\,,$$

(3.11a)

$$\tilde{\Gamma}_2 = (e_{12}C_1 + e_{22}C_2)\exp\left[\frac{1}{4}ik_0\int n_1^3(n_2^{-2} - n_1^{-2})\,ds_1\right]\,.$$

As a result, Eq.(3.28) takes the form

$$\mathcal{E}_e = \Phi_1 n_1^{-1/2}(C_1\mathbf{e}_1 + C_2\mathbf{e}_2)\exp(i\varphi_1),$$

(3.23a)

whereas Eqs. (3.31) become

$$\frac{dC_1}{ds_1} + p_{11}C_1 + p_{12}C_2 = 0,$$

(3.12a)

$$\frac{dC_2}{ds_1} + ik_0(n_1 - n_2)C_2 + p_{21}C_1 + p_{22}C_2 = 0.$$

In the case of an ordinary wave the indices 1 and 2 in the transform (3.11a) should be interchanged together with the replacement of $C_{1,2}$

by $C'_{1,2}$. This results in the following expression for the field:

$$\mathcal{E}_o = \Phi_2 n_2^{-1/2}(C'_1 \mathbf{e}_1 + C'_2 \mathbf{e}_2)\exp(i\varphi_2), \qquad (3.23b)$$

and equations for polarization components of the deformed wave

$$\frac{dC'_1}{ds_2} + ik_0(n_2 - n_1)C'_1 + p_{11}C'_1 + p_{12}C'_2 = 0,$$

$$(3.12b)$$

$$\frac{dC'_2}{ds_2} + p_{21}C'_1 + p_{22}C'_2 = 0.$$

Now one should make use of the seeds of the type of (3.17)–(3.19). The following set of equations then emerges for an extraordinary wave:

$$C_\alpha = C_\alpha^{(0)} + C_\alpha^{(1)} + C_\alpha^{(2)} + \cdots \qquad (\alpha = 1, 2); \qquad (3.17a)$$

$$C_1^{(0)}(s_{1in}) \neq 0, \qquad C_2^{(0)} \equiv 0; \qquad (3.18a)$$

$$\frac{dC_1^{(0)}}{ds_1} + p_{11}C_1^{(0)} = 0,$$

$$C_2^{(m)} = ik_0^{-1}(n_1 - n_2)^{-1}\left[\frac{dC_2^{(m-1)}}{ds_1} + p_{21}C_1^{(m-1)} + p_{22}C_2^{(m-1)}\right],$$

$$\frac{dC_1^{(m)}}{ds_1} + p_{11}C_1^{(m)} + p_{12}C_2^{(m)} = 0, \quad C_1^{(m)}(s_{1in}) \equiv 0 \quad (m \geq 1). \quad (3.19a)$$

Similarly, for an ordinary wave we have

$$C'_\alpha = C'^{(0)}_\alpha + C'^{(1)}_\alpha + C'^{(2)}_\alpha + \cdots \qquad (\alpha = 1, 2); \qquad (3.17b)$$

$$C'^{(0)}_2(s_{2in}) \neq 0, \qquad C'^{(0)}_1 \equiv 0; \qquad (3.18b)$$

$$\frac{dC'^{(0)}_2}{ds_2} + p_{22}C'^{(0)}_2 = 0,$$

$$C'^{(m)}_1 = ik_0^{-1}(n_2 - n_1)^{-1}\left[\frac{dC'^{(m-1)}_1}{ds_2} + p_{11}C'^{(m-1)}_1 + p_{12}C'^{(m-1)}_2\right],$$

$$\frac{dC'^{(m)}_2}{ds_2} + p_{21}C'^{(m)}_1 + p_{22}C'^{(m)}_2 = 0, \quad C'^{(m)}_2(s_{2in}) \equiv 0 \quad (m \geq 1).$$

$$(3.19b)$$

One should bear in mind that the coefficients p_{ij} in Eqs. (3.12a) and (3.17a)–(3.19a) and those in Eqs. (3.12b) and (3.17b)–(3.19b) differ to some extent since the former correspond to the extraordinary ray while the latter correspond to the ordinary one. An analogous remark refers to the polarization vectors e_1 and e_2 in Eqs. (3.23a) and (3.23b) as well as to the coefficients e_{ij} in the transform (3.11a) and in its counterpart for an ordinary wave.

3.3.3 ESTIMATIONS OF ACCURACY

That the QIA equations conform to the Budden equations for a plane layer gives a possible way for estimating the accuracy of the method in a reliable way, without recourse to the analysis of all residual terms omitted in Eqs. (3.33). Indeed, additional errors (with respect to these of the Budden method) in the method of split rays can only be associated with those terms in Eqs. (2.30) that correspond to the 3D inhomogeneity of the medium. But the errors they cause in the solution do not exceed μ_1 by an order of magnitude. Taking into account that the error of the Budden method scales as $\mu \sim k_0^{-1} l^{-1}$, we infer that for the region of relatively strong birefringence $\delta = \mu_1/\mu \gg 1$ the following estimate holds:

$$\left| \frac{\delta E}{E} \right| \lesssim \max(\mu, \mu_1) \,. \tag{3.32}$$

It can be expressed in a more uniform way

$$\max_{(s)} \left| \frac{\delta E}{E} \right| \lesssim \max\left[k_0^{-1} l_{\mathrm{B}}^{-1}, \max_{(s)} |\nu_{ij}(s)| \right] \,. \tag{3.33}$$

where l with the subscript B implies that the value of scale l is taken at the boundary of the region of interaction where $\delta \sim 1$.

In media possessing a strong anisotropy ($\mu_1 \sim 1$) the inequality (3.32) does not ensure the smallness of amplitude errors. Errors can however be reduced by matching a split solution with the independent normal waves of the Courant–Lax solutions. Errors introduced by the latter get smaller as μ_1 increases, while the QIA solutions exhibit the opposite behavior: their errors grow with the parameter of anisotropy μ_1. Therefore we may anticipate the existence of matching points on the rays which, when used, would minimize the error of the composite (matched) solution.

There also exists another technique for minimizing errors that underlies the "synthetic method" which combines the advantages both of the method of split rays and of the Courant–Lax method. As applied to electrodynamics, the synthetic approach will be presented below, in Section 3.4.

3.3.4   THE REGION OF LOCALIZATION OF RAY SPLITTING

The assessment of uniform accuracy (3.33) refers only to deformed normal waves and holds only in an interval of the ray where the transition from the region of relatively strong birefringence $\delta \gg 1$ to that of relatively weak birefringence $\delta \lesssim 1$ occurs only once. If there are multiple intersections of the regions with strong and weak birefringence, the complete geometric-optical solution to Maxwell's equations is to be composed of deformed normal waves defined in intervals of *monotonic* behavior of the parameter $\delta = \mu_1/\mu$.

The effective mutual transformation of extraordinary and ordinary waves and splitting of a ray into extraordinary and ordinary rays is localized in the region $\delta \sim 1$, i. e., in the region of moderate birefringence separating the regions with relatively strong, $\delta \gg 1$, and relatively weak, $\delta \lesssim 1$, birefringence (Budden, 1961; Ginzburg, 1970; Zheleznyakov, Kocharovskii and Kocharovskii, 1983).

Why the linear wave transformation occurs only in the region $\delta \lesssim 1$ can be explained in the following way. Within the region of weak birefringence, $\delta \ll 1$, the spatial scale of beats between polarization components $\Lambda = k^{-1}|n_1 - n_2|^{-1}$ substantially exceeds the scale l of medium inhomogeneities, so the beats of the scale Λ are simply not seen against the background of small-scale variations ($l \ll \Lambda$) in parameters of the medium. Noticeable wave and ray splitting does not occur under these conditions. On the other hand, in the region of relatively strong birefringence, $\delta \gg 1$, the waves $\mathcal{E}_1$ and $\mathcal{E}_2$ are almost independent and thus do not suffer mutual transformation. Hence, the mutual transformation of waves vanishes either for $\delta \ll 1$ or for $\delta \gg 1$ and is localized in the region $\delta \sim 1$. These considerations were already known to Budden (1961). The reader may find excellent illustrative examples in a work by Zheleznyakov, Kocharovskii and Kocharovskii (1983).

To sum up, the localization of mutual transformation between extraordinary and ordinary waves in the region $\delta \sim 1$ is explained by the competition of two factors. On the one hand, in the region $\delta \sim 1$ the difference in refractive indices is great enough for the beats between the waves not to disappear against the background of medium inhomogeneity. On the other hand, it is still small enough for waves to be capable of resonance both in frequency and wavelength.

3.3.5 MATCHING OF THE QIA SOLUTION WITH NORMAL WAVES

To describe the wave transformation in the region $\delta \sim 1$ Kravtsov (1968) suggested matching solutions of the type of non-interacting normal waves (2.15) with QIA solutions. The matching can be done appropriately at a point within the region of strong birefringence $\delta \gg 1$ which is still close to the interaction region $\delta \sim 1$. The details of that procedure were established by Naida (1974a).

The matching procedure becomes considerably simpler if one uses equations for the electric induction vector in the form for split rays. In that case, matching is implemented only in the regions of localization $\delta \sim 1$. On each occasion three deformed normal waves with the same value of wave vector $\mathbf{k}$ participate in the matching: a wave arriving at the matching point and two normal waves emerging from it. In this procedure, the relative accuracy of matched solutions is measured by $\mu_B \sim k_0^{-1} l_B^{-1}$ (see (3.33)), as is the case for a plane layer.

A linear transformation occurs at all points of local minimum of the parameter δ. If $\delta \gg 1$ in the region of the minimum, the linear transformation is rather weak there, $\sim \exp(-\delta)$. Therefore local extrema of the parameter δ in the region $\delta \gg 1$ should be ignored as a rule. On the contrary, the wave interaction in the region of relatively weak birefringence ($\delta \lesssim 1$) must be accounted for.

The procedure of matching the QIA solution with normal waves is elucidated by typical examples given below.

3.3.6 EXAMPLE 1. INCIDENCE OF A WAVE ON A BIREFRINGENT LAYER WITH A SINGLE MAXIMUM OF PARAMETER δ INSIDE THE LAYER

Let us assume that a superposition of ordinary and extraordinary waves with the same direction $\mathbf{t}^A$ of wave vectors is given at some initial point A (see Figure 3.1). The construction of the electromagnetic wave field in the vicinity of that point consists of the following steps.

(i) We draw extraordinary (1) and ordinary (2) rays through point A that correspond to the initial direction $\mathbf{t}^A$. Then we find phases φ_1 and φ_2 along these rays, with arbitrary initial values (at this stage) at the point A. Additionally, we construct the polarization unit vectors along each of the rays, $\mathbf{e}_1$ and $\mathbf{e}_2$, and $\mathbf{e}_1'$ and $\mathbf{e}_2'$, respectively, together with the amplitude functions Φ_1 and Φ_2 which satisfy the conservation law (3.29).

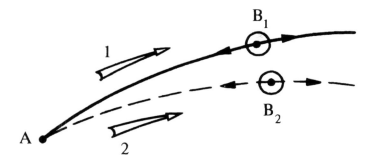

FIGURE 3.1. Explanations of the procedure employed in the split ray method to construct the full solution in the presence of a local maximum in $|n_1 - n_2|$. Solid and dashed lines show the extraordinary and ordinary rays, respectively. A is the initial point at which both rays correspond to the same wave vector direction. B_1 and B_2 are the points where $|n_1 - n_2|$ is at maximum. The iterative procedures are constructed in their vicinities. The arrows on the rays show the directions of integrations of Eqs. (3.12) (extraordinary seed) or (3.19) (ordinary seed); the arrows near the rays show the directions of wave propagation.

(ii) At each of the rays we find (at least approximately) the points B_1 and B_2 where the parameter $\delta = kl|n_1 - n_2|$ reaches a maximum.

(iii) In the region of strong birefringence $\delta \gg 1$, we construct iterative solutions for each ray in both the directions from points B_1 and B_2, using extraordinary or ordinary seeds (3.17a)–(3.19a) or (3.17b)–(3.19b), for example,

$$C_1^{(0)}(s_{1\text{in}}) = 1, \quad \text{or} \quad C_2^{(0)}(s_{1\text{in}}) = 1.$$

(iv) We employ the amplitudes $C_1(s_1)$, $C_2(s_1)$, $C_1'(s_2)$, and $C_2'(s_2)$ obtained by the iterative procedure as the initial condition for the system of equations (3.12a) for the extraordinary wave and for the analogous system of equations (3.12b) for the ordinary wave. Using these initial conditions the solutions $C_1(s_1)$ and $C_2(s_1)$ to the system (3.12a) are constructed on both sides of points B_1 and B_2. One then obtains the solutions $C_1'(s_2)$ and $C_2'(s_2)$ to the system (3.12b), and then the expressions for corresponding fields $\mathcal{E}_e(s_1)$ and $\mathcal{E}_o(s_2)$ from Eqs. (3.23a) and (3.23b).

(v) One looks for the coefficients $K_{1,2}$ of the initial field decomposition

$$\mathcal{E}^A = K_1 \mathcal{E}_e(A) + K_2 \mathcal{E}_o(A) \, . \tag{3.34}$$

The solvability of this equation up to terms $\lesssim k_0^{-1} l^{-1}$ is ensured by both sides of (3.34) being transverse to $\mathbf{t}^A$ upon action of the tensor $\widehat{\varepsilon}$.

(vi) The superposition

$$\mathcal{E}^A = K_1 \mathcal{E}_e(s_1) + K_2 \mathcal{E}_o(s_2)$$

of two deformed normal waves,

$$K_1 \mathcal{E}_e(s_1) \qquad \text{and} \qquad K_2 \mathcal{E}_o(s_2) \, ,$$

with each being localized on the respective ray, gives the desired solution of the geometric-optical problem.

3.3.7   EXAMPLE 2. OBSERVATION OF A POINT SOURCE THROUGH A BIREFRINGENT LAYER WITH A SINGLE MAXIMUM IN THE PARAMETER δ

Assume that a point source is located at the point A. This implies that the directions $\mathbf{t}^A$ of wave vectors and the initial values $\mathcal{E}^A$ of fields are specified on some small sphere S^A surrounding the point A. We want to find the field $\mathcal{E}$ at a given point B (see Figure 3.2).

In this case, the solution also includes one preliminary step in addition to six steps listed above in example 1: we should find two rays $\widehat{A\,B}$ of types 1 and 2 which would join the source A with the observation point B. For each of the rays the points of intersection A_1 and A_2 with the initial sphere S^A are to be determined. Then the calculation breaks into two parts corresponding to the rays $\widehat{A_1 B}$ and $\widehat{A_2 B}$. For each of them the statement of the problem is essentially that of example 1. Indeed, for the ray $\widehat{A_1 B}$, the initial direction of wave-vectors, $\mathbf{t}^{A1}$, and the initial field, $\mathcal{E}^{A1}$, are given at the point A_1. Now, in accordance with step (i) of example 1, a ray with the same direction of the wave-vector $\mathbf{t}^{A1}$, but corresponding to a normal wave of type 2, should be drawn through the point A_1. In the same manner, an additional ray is to be drawn through the point A_2, but now corresponding to a normal wave of type 1. Clearly, these additional rays do not reach the point B. However they are necessary to determine the components of the initial fields $\mathcal{E}^{A1}$ and $\mathcal{E}^{A2}$ which correspond to rays $\widehat{A_1 B}$ and $\widehat{A_2 B}$, in accordance with step (v) and (vi).

What follows is clear from Figure 3.2. At the point B we obtain the sum of two waves arriving by different rays $\widehat{A_1 B}$ and $\widehat{A_2 B}$. This implies

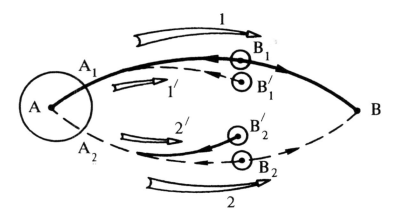

FIGURE 3.2. Radiation due to a point source (surrounded by a small sphere S^A in a birefringent layer with a local maximum in $|n_1 - n_2|$. 1 and 2 are the basic rays reaching the observation point B; $1'$ and $2'$ are the auxiliary rays, and $B_{1,2}$ and $B'_{1,2}$ are the points at which $|n_1 - n_2|$ attains a maximum. The arrows on the rays show the directions of integration in Eqs. (3.12) or (3.16) while those near the rays show the wave propagation directions.

that two sources will be seen from the point B, one in polarization 1, and the other in polarization 2. Equations of the type of (3.12a) or (3.12b) for ordinary and extraordinary waves need to be solved at all four rays that appear in the treatment of the problem. For two basic rays $\widehat{A_1 B}$ and $\widehat{A_2 B}$, seed iterations are constructed on both sides of the point with maximum δ, whereas they are constructed only in the direction to the source A for two additional rays.

3.3.8   EXAMPLE 3. THE INCIDENCE OF A WAVE ON A BIREFRINGENT LAYER WITH TWO MAXIMA AND ONE ESSENTIAL MINIMUM OF THE PARAMETER δ

The example is explained in Figure 3.3. When speaking of an essential minimum we imply that at that point the parameter δ does not exceeds unity: $\delta_{\min} \lesssim 1$ on the order of value.

The sequence of operations is clear in this case from Figure 3.3. Finally, an original ray splits into four rays, two of type 1 and two of type 2.

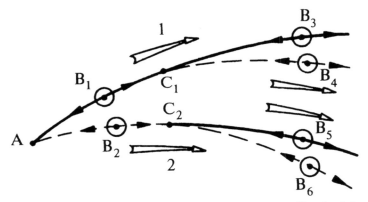

FIGURE 3.3. Construction of the full solution in the presence of local minima of $|n_1 - n_2|$. C_1 and C_2 are the the points of minima where the rays split and the solutions are matched. B_1, B_2, ... are the points of local maxima. The remaining notations are as in Figures 3.1 and 3.2.

3.3.9   Example 4. Observation of a point source through a birefringent layer with two maxima and one essential minimum in the parameter δ

This example (see Figure 3.4) stands in the same relation to example 3, as example 2 does to 1. The sequence of steps is apparent from Figure 3.4. A new element (compared to example 3) here is the construction of rays joining points A and B. The corresponding procedure implies some ray selection. We assume that the rays of type 1 leave point A in different directions with a plausible angular separation between them. Let us find a point of local extremum for the parameter δ at each of them. We calculate the direction $\mathbf{t}$ of the wave vector at each point of minimum, and draw additional rays of type 2 in this direction. A similar procedure is implemented for the rays of type 2 except for drawing additional rays of type 1 through local minimum points of the parameter δ.

Therefore, four families of rays emanate from the point A:

(1) original rays of type 1;

(2) composite rays of type $1 \to 2$ (emitted with index 1, but extended with index 2);

(3) composite rays of type $2 \to 1$ (emitted with index 2, but extended with index 1);

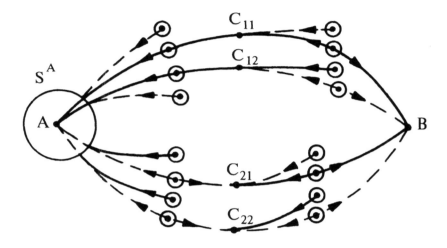

FIGURE 3.4. Observations of a point source A through a birefringent layer with minima of $|n_1 - n_2|$ present on each of the rays. Four rays of types $1 \to 1$, $1 \to 2$, $2 \to 1$, and $2 \to 2$ arrive to the point of observation. C_{11}, C_{12}, C_{21}, and C_{22} denote the points of local minima of $|n_1 - n_2|$ where the rays split. The remaining notations are of Figures 3.1 and 3.2.

(4) original rays of type 2.

To continue, among the rays of each type one should find the ray which would join points A and B. Finally, the wave fields are to be determined along each of these rays, as in example 2, by applying the procedure of example 3.

The net result is elucidated by Figure 3.4. As a result, the field at the point B is expressed as a sum of four waves arriving along different rays from different directions. Accordingly, at this point an observer will see four images of the source A, two in polarization 1 and two in polarization 2. Equations of the type of (3.12a) or (3.12b) are to be solved (in order to construct deformed normal waves) in 16 different ray intervals. In eight of them, solutions are constructed on one side of the region of seed iteration, while in other eight, solutions are sought on both sides.

3.3.10 CRITERIA OF DISTINGUISHABILITY OF SPLIT RAYS

It is known that the rays have a dual nature. On the one hand, a ray is a mathematical object, i. e. an infinitely thin line in space, in which in the approximation of geometrical optics the wave field is fitted. On the other hand, it is a physical object whose parameters, say, the thickness, can be measured. The physical aspects of the concept of a ray have been discussed by Kravtsov and Orlov (1980, 1990), and also by Kravtsov (1988). In line with that concept, a physical ray is associated with the Fresnel volume surrounding the ray whereas physical distinguishability of rays implies the possibility of separating the Fresnel volumes (a Fresnel volume is the union of all first Fresnel zones associated the ray).

Similar distinguishability criteria should also hold in an anisotropic medium: as soon as the Fresnel volumes of rays cease to intersect, the rays can be regarded as existing on their own, i.e. as admitting a distinguishability by means of physical instruments. These include orifices, slits, lenses, reflectors, antennae, phase arrays, etc.

In an anisotropic medium, the list of instruments can be supplemented by polarizers performing the selection of polarization, and by travelling wave antennae discriminating between ordinary and extraordinary waves by their phase velocity.

Needless to say, in the interaction region, $\delta \lesssim 1$, the split rays cannot be distinguished by any physical instrument. It seems to us that the question of the distinguishability of rays leaving the interaction region in a weakly anisotropic medium calls for further detailed investigation.

3.4 Generalization of the method of split rays to strongly anisotropic media: synthetic approach

3.4.1 INTRODUCTORY REMARKS

The method described above of split rays requires anisotropy to be weak, $\mu_1 \ll 1$. Meanwhile one can lift this constraint by using the original, unsimplified, normal waves without resorting to the expansions of refractive indices and polarization vectors in a small anisotropy parameter μ_1.

A corresponding modification of the method of split rays was proposed by Naida (1977a) and received additional mathematical justification in a work by Naida and Prudkovskii (1977). Here, this modification is outlined under the name of the "*synthetic approach*", since it combines the advantages of the QIA in the form of split rays, which is capable of describing the interaction of waves in 3D inhomogeneous media but fails in the case of strongly anisotropic media with $\mu_1 \sim 1$, with the advantages of the Courant–Lax method, which also applies to 3D inhomogeneous media, and allows for a strong anisotropy. Notice that, taken without modifications, the Courant–Lax method fails to describe the transformation of normal waves.

One may also argue that the "synthetic approach" is the generalization of the Budden method to the case of a 3D inhomogeneous medium.

With this synthesis of the QIA and Courant–Lax methods, a new useful feature emerges, namely, the synthetic approach suggests the possibility of describing the interaction of normal waves in the vicinities of peculiar, "degenerate" directions of the wave vector $\mathbf{k}$ in a strongly anisotropic medium. The refractive indices of two normal waves coincide in such directions: $n_1(\mathbf{r}, \mathbf{k}) = n_2(\mathbf{r}, \mathbf{k})$. This favors a strong interaction between waves with near-parallel phase fronts. The first, relatively straightforward attempt to construct the theory of interaction in the vicinities of degenerate directions (the "quasi-degenerate" approximation of geometrical optics suggested by Apresyan, Kravtsov, Yashin, and Yashnow (1976)) was not fully successful. Here we present a more rigorous approach which allows for ray splitting within the region of interaction.

Equations for interacting waves in the framework of the synthetic approach could be written by analogy with Section 3.3, with the replacement of simplified normal waves by their original counterparts. We postpone this till the end of the present chapter, and write the equation in Section 3.4 after presenting the synthetic approach as applied to vector wave fields of an arbitrary nature. In that way we prepare the ground for the generalization of the quasi-isotropic approximation to elastic waves (Chapter 6) as well as to the de Broglie waves for spin particles described by a spinor wave function (Section 7).

We notice that the synthetic method was initially suggested as a natural modification of the Courant–Lax method, while the method of split rays within the QIA was formulated as a subasymptotics of the synthetic method under weak anisotropy (Naida, 1977a). Therefore, the derivation of the synthetic approach simultaneously serves as an additional justification of the method of split rays. It is for that reason that we attempt to present a full (although concise) derivation here.

3.4.2 MODIFIED FUNDAMENTAL IDENTITY

The synthetic approach relies on a modified (revisited) fundamental identity of the Courant–Lax approach. This identity provides grouping of partial derivatives in full derivatives along the ray. In the case of Maxwell's equations it is given by (2.43).

We begin with an arbitrary linear system of equations,

$$L\left(t, \mathbf{r}; \frac{\partial}{\partial t}, -i\nabla\right)\mathcal{X} = 0, \qquad (3.35)$$

without imposing any restrictions on the number of components of the vector $\mathcal{X}$ and the order of equations in partial derivatives. Aside from Maxwell's equations, the equations of elastic waves, the Pauli and Dirac equations, and a few other equations of similar type can be cast in the form (3.35).

The function L in (3.35) is assumed to be a local function of its variables, i. e. the operator L is purely differential, although the ultimate result also applies to integro-differential operators encountered, say, in electrodynamics in the presence of spatial dispersion.

We take advantage of the ordinary geometric-optical substitution

$$\mathcal{X}(t, \mathbf{r}) = \mathbf{X}(t, \mathbf{r}) \exp[i\varphi(t, \mathbf{r})], \qquad (3.36)$$

which reduces Eq. (3.35) to the form

$$L(t, \mathbf{r}; \omega, \mathbf{k})\mathbf{X} + \tilde{L}[\mathbf{X}] = \delta L(\mathbf{X}), \qquad (3.37)$$

where $\omega(t, \mathbf{r}) = -\partial\varphi/\partial t$, $\mathbf{k}(t, \mathbf{r}) = \nabla\varphi$, and the symbol $L(t, \mathbf{r}; \omega, \mathbf{k})$ implies the same function as in (3.35), but in ordinary variables (as opposed to operator ones). The operator $\tilde{L}$ in (3.37) is given by

$$\tilde{L}[\mathbf{X}] = i\left[\left(\frac{\partial L}{\partial \Omega}\frac{\partial}{\partial t} - \frac{\partial L}{\partial \mathbf{k}}\nabla\right)\right.$$

$$\left. + \frac{1}{2}\left(\frac{\partial \omega}{\partial t}\frac{\partial^2 L}{\partial \omega^2} - 2\nabla\omega\frac{\partial^2 L}{\partial \Omega \partial \mathbf{k}} - \sum_{m,n=1}^{3}\frac{\partial k_m}{\partial x_n}\frac{\partial^2 L}{\partial k_m \partial k_n}\right)\right]\mathbf{X}. \qquad (3.38)$$

Similar expressions were obtained in many works on asymptotic methods of wave theory for dispersive media; see, for example, Maslov and Fedoryuk (1981) and Kravtsov and Orlov (1990).

The remainder term $\delta L[\mathbf{X}]$ is of the order of

$$|\mathbf{X}|^{-1}\mathcal{O}\left(|\mathbf{k}|^{-1}l\left|\frac{\partial^2\mathbf{X}}{\partial x^2}\right|\right)$$

with respect to $\tilde{L}[\mathbf{X}]$.

We assume that for the solutions $\mathbf{X}(x)$ considered here

$$\left|\frac{\partial^2\mathbf{X}}{\partial x^2}\right|\sim l^{-2}$$

far from the interaction region. Therefore the function $\delta L[\mathbf{X}]$ there is of the order of the small geometric-optical parameter $\mu\sim(kl)^{-1}$ as compared to $\tilde{L}[\mathbf{X}]$. Corresponding solutions $\mathbf{X}(x,t)$ are the non-interacting (generally speaking, deformed) normal waves. In the interaction region the term $\delta L[\mathbf{X}]$ increases somewhat because of intermodal beats, but, as is shown in additional analysis, it nevertheless remains small (of order μ) and will not play an essential role in what follows.

The next step of the geometric-optical method is the choice of eikonal φ in the substitution (3.36). In the framework of the Courant–Lax method the eikonal $\varphi(t,\mathbf{r})$ associated with a mode with number α obeys the equation

$$\frac{\partial\varphi_\alpha}{\partial t}=-\Omega_\alpha(t,\mathbf{r},\nabla\varphi_\alpha),$$

where Ω_α is one of the solutions of the dispersion equation

$$\det L(t,\mathbf{r};\Omega_\alpha,\mathbf{k})=0,\qquad\alpha=1,2,\ldots,M.\qquad(3.39)$$

An initial condition should be specified for the eikonal φ_α, for example, in the form $\varphi_\alpha(t_{\mathrm{in}},\mathbf{r})=\varphi_{\mathrm{in}}(\mathbf{r})$ at $t=t_{\mathrm{in}}$.

We relate the eikonal φ_α with the system of rays $\mathbf{r}=\mathbf{r}_\alpha(t;\mathbf{r}_{\mathrm{in}})$ obeying the Hamiltonian system of equations

$$\frac{d\mathbf{r}}{dt}=\frac{\partial\Omega_\alpha(t,\mathbf{r},\mathbf{k})}{\partial\mathbf{k}},\qquad\frac{d\mathbf{k}}{dt}=-\frac{\partial\Omega_\alpha(t,\mathbf{r},\mathbf{k})}{\partial\mathbf{r}}\qquad(3.40)$$

with initial conditions

$$\mathbf{r}(t_{\mathrm{in}};\mathbf{r}_{\mathrm{in}})=\mathbf{r}_{\mathrm{in}}\quad\text{and}\quad\mathbf{k}(t_{\mathrm{in}};\mathbf{r}_{\mathrm{in}})=\frac{\partial\varphi_{\mathrm{in}}}{\partial\mathbf{r}_{\mathrm{in}}}\qquad(3.41)$$

for an initial point and initial wave vector, respectively. For the rays subject to (3.40), we obtain

$$\frac{d\varphi_\alpha}{dt} = \mathbf{k}_\alpha \frac{d\mathbf{r}_\alpha}{dt}, \qquad \frac{d\omega_\alpha}{dt} = \frac{\partial \Omega_\alpha(t, \mathbf{r}, \mathbf{k})}{\partial t}. \tag{3.42}$$

In contrast with the Courant–Lax method which employs the eigenvectors of matrix L, we will use the polarization null-vectors $e_\alpha(t, \mathbf{r}, \mathbf{k})$ defined as

$$L[t, \mathbf{r}; \Omega_\alpha(t, \mathbf{r}, \mathbf{k}), \mathbf{k}]e_\alpha(t, \mathbf{r}, \mathbf{k}) = 0 \quad (\alpha = 1, \cdots, M). \tag{3.43}$$

It is assumed that the homogeneous system of equations (3.43) admits M linearly independent solutions e_α which belong to different eigenvalues $\Omega_\alpha(t, \mathbf{r}, \mathbf{k})$ with account for their multiplicity. If all the roots $\Omega_\alpha(t, \mathbf{r}, \mathbf{k})$ of the dispersion equation (3.39) are different (non-degenerate), the number of eigenvectors e_α coincides with that of roots. The multiple roots are enumerated in such a way as to correspond to different linearly independent null vectors e_α. For instance, if $\Omega = \Omega(t, \mathbf{r}, \mathbf{k})$ is a double root of Eq. (3.39), then it should be associated with two indices, for example, $\alpha = 1$ and $\alpha = 2$ and two different linearly independent null-vectors e_1 and e_2.

Left null-vectors will be replaced by the vectors $e_\alpha^T(t, \mathbf{r}, \mathbf{k})$, related to the transposed matrix L^T:

$$L^T[t, \mathbf{r}; \Omega_\alpha(t, \mathbf{r}, \mathbf{k}), \mathbf{k}]e_\alpha^T(t, \mathbf{r}, \mathbf{k}) = 0 \quad (\alpha = 1, \ldots, M). \tag{3.44}$$

For a Hermitian matrix L the vectors $e_\alpha^T(t, \mathbf{r}, \mathbf{k})$ and $e_\alpha^*(t, \mathbf{r}, \mathbf{k})$ are collinear.

At polarization degeneration, when $\Omega_\alpha(t, \mathbf{r}, \mathbf{k}) = \Omega_\beta(t, \mathbf{r}, \mathbf{k})$, we impose the following orthonormality condition on vectors e_α and e_β^T:

$$\left(e_\beta^T(t, \mathbf{r}, \mathbf{k}), \frac{\partial L(t, \mathbf{r}; \Omega, \mathbf{k})}{\partial \Omega} \bigg|_{\Omega = \Omega_\alpha(t, \mathbf{r}, \mathbf{k})} e_\alpha(t, \mathbf{r}, \mathbf{k}) \right) = \delta_{\alpha\beta}. \tag{3.45}$$

If the matrix L has the structure

$$L(t, \mathbf{r}; \Omega, \mathbf{k}) = \Omega^N B(t, \mathbf{r}) + A(t, \mathbf{r}; \mathbf{k}), \tag{3.46}$$

typical for equations of mathematical physics, then, at $\Omega_\alpha = \Omega_\beta$ the condition (3.45) holds identically, while the vectors e_α and e_α^T introduced in manner outlined above satisfy the identity

$$\left(\mathbf{e}_\beta^{\mathrm{T}}(t,\mathbf{r};\mathbf{k}),\, L[t,\mathbf{r};\Omega_\alpha(t,\mathbf{r};\mathbf{k}),\mathbf{k}]\boldsymbol{\xi}\right) = \left[\Omega_\alpha^N(t,\mathbf{r};\mathbf{k}) - \Omega_\beta^N(t,\mathbf{r};\mathbf{k})\right]$$

$$\times N^{-1}\Omega_\alpha^{1-N}(t,\mathbf{r};\mathbf{k})\left(\mathbf{e}_\beta^{\mathrm{T}}(t,\mathbf{r};\mathbf{k}),\, \frac{\partial L(t,\mathbf{r};\Omega,\mathbf{k})}{\partial\Omega}\bigg|_{\Omega=\Omega_\alpha(t,\mathbf{r},\mathbf{k})}\boldsymbol{\xi}\right),$$

$$\tag{3.47}$$

where $\boldsymbol{\xi}$ is an arbitrary M-component vector whereas N is the same integer as in (3.46). From this, *the modified fundamental identity* follows, which is valid for arbitrary α,β (Naida, 1977a):

$$\left(\mathbf{e}_\beta^{\mathrm{T}}(t,\mathbf{r};\mathbf{k}),\, \frac{\partial L(t,\mathbf{r};\Omega,\mathbf{k})}{\partial\Omega}\bigg|_{\Omega=\Omega_\alpha(t,\mathbf{r},\mathbf{k})}\mathbf{e}_\alpha(t,\mathbf{r};\mathbf{k})\right)$$

$$= -\frac{\partial\Omega_\alpha(t,\mathbf{r},\mathbf{k})}{\partial\mathbf{k}}\delta_{\alpha\beta}$$

$$+[\Omega_\beta^N(t,\mathbf{r};\mathbf{k}) - \Omega_\alpha^N(t,\mathbf{r};\mathbf{k})]N^{-1}\Omega_\alpha^{1-N}(t,\mathbf{r};\mathbf{k}) \tag{3.48}$$

$$\times\left(\mathbf{e}_\beta^{\mathrm{T}}(t,\mathbf{r};\mathbf{k}),\, \frac{\partial L(t,\mathbf{r};\Omega,\mathbf{k})}{\partial\Omega}\bigg|_{\Omega=\Omega_\alpha(t,\mathbf{r},\mathbf{k})}\frac{\partial\mathbf{e}_\beta(t,\mathbf{r};\mathbf{k})}{\partial\mathbf{k}}\right).$$

For $\Omega_\beta(t,\mathbf{r};\mathbf{k}) = \Omega_\alpha(t,\mathbf{r};\mathbf{k})$ this identity reduces to the fundamental identity of Courant and Lax:

$$\left(\mathbf{e}_\beta^{\mathrm{T}}(t,\mathbf{r};\mathbf{k}),\, \frac{\partial L(t,\mathbf{r};\Omega,\mathbf{k})}{\partial\Omega}\bigg|_{\Omega=\Omega_\alpha(t,\mathbf{r},\mathbf{k})}\mathbf{e}_\alpha(t,\mathbf{r};\mathbf{k})\right) = -\frac{\partial\Omega_\alpha(t,\mathbf{r},\mathbf{k})}{\partial\mathbf{k}}\delta_{\alpha\beta}.$$

$$\tag{3.48a}$$

With the help of identities (3.47) and (3.48) one can readily cast the initial system of equations (3.35) (or (3.37)) in a form convenient for a conversion to geometrical optics.

Let us resolve the sought-for solution $\mathbf{X}(x)$ into a superposition of all M linearly-independent polarization vectors:

$$\mathbf{X} = C_1\mathbf{e}_1(t,\mathbf{r},\mathbf{k}_1) + \cdots + C_M\mathbf{e}_M(t,\mathbf{r},\mathbf{k}_1). \tag{3.49}$$

All polarization vectors $\mathbf{e}_\alpha$ refer to the same system of rays (here, to rays with polarization index 1).

Assume that among all frequencies Ω_α only Ω_1 and Ω_2 are close to each other. Then, substitution of (3.49) into (3.37) and contraction of the resultant expression with $e_1^T(t, \mathbf{r}; \mathbf{k}_1), \cdots, e_M^T(t, \mathbf{r}; \mathbf{k}_1)$ yields the following two equations for amplitudes C_1 and C_2:

$$\frac{dC_1}{dt_1} + P_{11}^1 C_1 + P_{12}^1 C_2 = \xi_1[\mathbf{X}], \tag{3.50}$$

$$\frac{dC_2}{dt_1} + i(\Omega_2 - \Omega_1)C_2 + P_{21}^1 C_1 + P_{22}^1 C_2 = \xi_2[\mathbf{X}], \tag{3.51}$$

and $M - 2$ equations for all other amplitudes

$$C_\gamma = (\Omega_\gamma^N - \Omega_1^N)^{-1} N \Omega_\gamma^{N-1} (e_\gamma^T, \tilde{L}[\mathbf{X}] + \delta L[\mathbf{X}]) \quad (\gamma = 3, \ldots, M). \tag{3.52}$$

Here we denote:

$$\xi_1[\mathbf{X}] = \frac{\Omega_1^N - \Omega_2^N}{N \Omega_1^{N-1}} \left(e_1^T, \frac{\partial L}{\partial \Omega} \bigg|_{\Omega = \Omega_1} \frac{\partial e_2}{\partial \mathbf{k}} \right) \nabla C_2 - i(e_1^T, \delta \tilde{L}[\mathbf{X}]),$$

$$\xi_2[\mathbf{X}] = \frac{\partial(\Omega_1 - \Omega_2)}{\partial \mathbf{k}} \nabla C_2$$

$$+ \frac{\Omega_2^N - \Omega_2^N}{N \Omega_1^{N-1}} \left(e_2^T, \frac{\partial L}{\partial \Omega} \bigg|_{\Omega = \Omega_1} \frac{\partial e_1}{\partial \mathbf{k}} \right) \nabla C_1$$

$$+ i \left[(\Omega_2 - \Omega_1) - \frac{\Omega_2^N - \Omega_2^N}{N \Omega_1^{N-1}} \right] C_2 - i(e_2^T, \delta \tilde{L}[\mathbf{X}]), \tag{3.53}$$

$$\delta \tilde{L}[\mathbf{X}] = \delta L[\mathbf{X}] + \sum_{\gamma=3}^M \left(i \frac{\partial L}{\partial \mathbf{k}} e_\gamma \nabla C_\gamma - \tilde{L}[e_\gamma] C_\gamma \right).$$

The designation d/dt_1 in Eqs. (3.50), (3.51), as in the Courant–Lax transfer equations, implies the full derivative with respect to t along the ray (3.40) with a polarization index $\alpha = 1$. $P_{\alpha\beta}^\gamma$ in Eqs. (3.50), (3.51) stands for

$$P_{\alpha\beta}^\gamma = -i \left(e_\alpha^T(t, \mathbf{r}, \mathbf{k}_\gamma), \tilde{L}[e_\beta^T(t, \mathbf{r}, \mathbf{k}_\gamma)] \right). \tag{3.54}$$

3.4.3 SYNTHETIC EQUATIONS AND TWO-SIDED ASYMPTOTICS

Eqs. (3.50)-(3.52) yield an equivalent representation of Eqs. (3.35). This representation is only worthwhile in the case when the frequency $\omega_1 = \Omega_1(t, \mathbf{r}, \mathbf{k}_1)$ is approached by a single frequency from all the remaining Ω_α, say by $\Omega_2(t, \mathbf{r}, \mathbf{k}_1)$. Only in this case all components C_γ with $\gamma > 2$ can be assumed to be small. Eqs. (3.50)-(3.53) comprise four groups of contributions: (i) the terms with full derivatives along the ray, i. e., with $d/dt_1 = \partial/\partial t + \mathbf{v}_{1g} \nabla$ ($\mathbf{v}_{1g}$ is the wave group velocity); (ii) combinations including the difference $\Omega_1 - \Omega_2$ that tends to zero as $\Omega_2 \to \Omega_1$; (iii) small terms of the order of $\mu \sim (kl)^{-1}$, denoted as δL; (iv) terms with small amplitudes $C_3, \ldots, C_M$ which only participate weakly in the interaction. The presence of terms of (ii), (iii), and (iv) types renders the right-hand sides of Eqs. (3.50)-(3.52) sufficiently small, at least, as $\Omega_2 \to \Omega_1$. This allows us to choose the zeroth-order approximation $\mathbf{X}$ by the formulae

$$\mathbf{X} = C_1 \mathbf{e}_1(t, \mathbf{r}, \mathbf{k}_1) + C_2 \mathbf{e}_2(t, \mathbf{r}, \mathbf{k}_1),$$

$$C_\gamma = 0 \qquad (3 \leq \gamma \leq M),$$

$$(3.55)$$

$$\frac{dC_\alpha}{dt_1} + i(\Omega_\alpha - \Omega_1)C_\alpha + \sum_{\beta=1}^{2} P_{\alpha\beta}^1 C_\beta = 0 \quad (\alpha = 1, 2). \tag{3.56}$$

One can easily rewrite (3.56) in a form involving spatial derivatives along the ray with index 1 instead of temporal derivatives:

$$\frac{dC_\alpha}{ds_1} + i \left| \frac{\partial \Omega_1}{\partial \mathbf{k}} \right|^{-1} (\Omega_\alpha - \Omega_1)C_\alpha + \sum_{\beta=1}^{2} p_{\alpha\beta}^1 C_\beta = 0 \quad (\alpha = 1, 2). \tag{3.57}$$

where

$$p_{\alpha\beta}^1 = \left| \frac{\partial \Omega_1}{\partial \mathbf{k}} \right|^{-1} P_{\alpha\beta}^1 = -i \left| \frac{\partial \Omega_1}{\partial \mathbf{k}} \right|^{-1} (\mathbf{e}_\alpha^T, \tilde{L}[\mathbf{e}_\beta]) \qquad (\mathbf{k} = \mathbf{k}_1). \tag{3.58}$$

In the region of strong birefringence, $\delta \gg 1$ the equation admits a solution in the form of a "weakly deformed normal wave":

$$C_\alpha = C_\alpha^{(0)} + C_\alpha^{(1)} + C_\alpha^{(2)} + \cdots \quad (\alpha = 1, 2), \tag{3.59}$$

in which the expansion is carried out in inverse powers of δ, i.e., $C_\alpha^{(m)} \sim \delta^{-m}$. In the zeroth approximation $C_2^{(0)} = 0$ and

$$\frac{dC_1^{(0)}}{dt} + P_{11}^1 C_1^{(0)} = 0, \qquad |C_1^{(0)}(t_{\text{in}})| \sim 1. \tag{3.60}$$

Subsequent terms of the series (3.59) satisfy the equations

$$C_2^{(1)} = \frac{i P_{21}^1 C_1^{(0)}}{\Omega_2 - \Omega_1},$$

$$\frac{dC_1^{(1)}}{dt} + P_{11}^1 C_1^{(1)} = -P_{12}^1 C_2^{(1)}, \tag{3.61}$$

$$C_1^{(0)}(t_{\text{in}}) = 0;$$

$$C_2^{(m)} = \frac{i}{\Omega_2 - \Omega_1} \left(\frac{dC_2^{(m-1)}}{dt} + P_{21}^1 C_1^{(m-1)} + P_{22}^1 C_2^{(m-1)} \right), \tag{3.62}$$

$$\frac{dC_1^{(m)}}{dt} + P_{11}^1 C_1^{(m)} = -P_{12}^1 C_2^{(m)},$$

$$C_1^{(m)}(t_{\text{in}}) = 0; \qquad m \geq 2.$$

The zeroth approximation equation (3.60) coincides exactly with the zeroth approximation equation (2.42a) of the Courant–Lax method. In addition, in the absence of birefringence, Eqs. (3.56) reduce to equations that are analogous to the Rytov geometric-optical equations for an isotropic medium.

To conclude, we may argue that Eqs. (3.56) allow for a "transition from both sides": to the region of strong birefringence, $\delta \gg 1$, and to the region of negligibly small birefringence, $\delta \to 0$. Thus one can expect these equations to ensure an improved accuracy of approximation. Whether this improved accuracy is achieved can be clarified by resorting to the method devised by Naida (1977a) and described in Chapter 2 as applied to the QIA. A detailed analysis of the accuracy of the "synthetic" method was performed by Naida and Prudkovskii (1977).

By providing the two-side asymptotics, the synthetic approach incorporates as a subasymptotics the quasi-isotropic approximation in the form of equations for the split rays. This peculiar feature of the synthetic approach will be illustrated below in Sections 3.4.4 and 3.4.5 as applied to electrodynamics.

3.4.4 SYNTHETIC METHOD IN THE ELECTRODYNAMICS OF 3D INHOMOGENEOUS ANISOTROPIC MEDIUM

We shall illustrate the synthetic method in the example of Maxwell's equations for an inhomogeneous anisotropic medium which we take in the form

$$\frac{\partial \widehat{\varepsilon} \mathcal{E}}{\partial t} - c\,\mathrm{curl}\,\mathcal{H} = 0, \quad \frac{\partial \mathcal{H}}{\partial t} + c\,\mathrm{curl}\,\mathcal{E} = 0. \tag{3.63}$$

System (3.63) can be written as (3.35) provided $\mathcal{X}$ is a six-component vector $\mathcal{X} = \{\mathcal{E}, \mathcal{H}\}$ and L is 6×6 matrix containing spatial derivatives. Then it will turn out that the polarization vectors $\{\mathbf{e}_a, \mathbf{h}_a\}$ are to be determined from the ordinary relations

$$c^{-2}\Omega_a^2 \widehat{\varepsilon}\mathbf{e}_a + \mathbf{k} \times \mathbf{k} \times \mathbf{e}_a = 0, \quad \mathbf{h}_a = c\Omega_a^{-1}\mathbf{k} \times \mathbf{e}_a,$$

$$\mathbf{e}_a^{\mathrm{T}} = \mathbf{e}_a^*, \quad \mathbf{h}_a^{\mathrm{T}} = \mathbf{h}_a^*, \tag{3.64}$$

while the fields $\mathcal{E}$ and $\mathcal{H}$ take the form

$$\mathcal{E}(t, \mathbf{r}) = (C_1 \mathbf{e}_1 + C_2 \mathbf{e}_2)\, e^{i\varphi_1}, \quad \mathcal{H}(t, \mathbf{r}) = (C_1 \mathbf{h}_1 + C_2 \mathbf{h}_2) e^{i\varphi_1}. \tag{3.65}$$

We obtain the expressions for coefficients p_{ab}^1 in (3.57), according to (3.58),

$$p_{ab}^1 = \left|\frac{\partial \Omega_1}{\partial \mathbf{k}}\right|^{-1} \left[\mathbf{e}_a^*\left(\frac{\partial \widehat{\varepsilon}\mathbf{e}_b}{\partial t} - c\,\mathrm{curl}\,\mathbf{h}_b\right) + \mathbf{h}_a^*\left(\frac{\partial \mathbf{h}_b}{\partial t} + c\,\mathrm{curl}\,e_b\right)\right]. \tag{3.66}$$

Equation (3.57) gives an approximation with an overall accuracy $\mu \sim |\mathbf{k}|^{-1}l_{\mathrm{B}}^{-1}$, where l_{B}^{-1} is the inhomogeneity scale at the boundary of the interaction region $\delta \sim 1$. Introducing the refractive indices $n_a(t, \mathbf{r}, \mathbf{k}) = |\mathbf{k}|c\Omega_a^{-1}(t, \mathbf{r}, \mathbf{k})$, we can rewrite the coefficients p_{ab}^1 in the form

$$p_{ab}^1 = |\beta_1|^{-1} \left[\mathbf{e}_a^*\left(\frac{\partial \widehat{\varepsilon}\mathbf{e}_b}{c\partial t} - \mathrm{curl}\,\mathbf{h}_b\right)\right.$$

$$\left. + \mathbf{h}_a^*\left(\frac{\partial \mathbf{h}_b}{c\partial t} + \mathrm{curl}\,e_b\right)\right]. \tag{3.67}$$

Accordingly, Eq. (3.57) is modified to:

$$\frac{dC_a}{ds_1} + ik_0 \, |\beta_1|^{-1} \left(\frac{n_1}{n_a} - 1 \right) C_a + \sum_{b=1}^{2} p_{ab}^1 C_b = 0 \quad (a = 1, 2). \quad (3.68)$$

Here and in (3.67) β_1 is defined according to (2.11):

$$\beta_1 = \frac{\partial |\mathbf{k}| n_1^{-1}}{\partial \mathbf{k}}.$$

Equations (3.68) with the coefficients $p_{\alpha\beta}^1$ from (3.66) or (3.67) define the desired synthetic asymptotic solutions of Maxwell's equations (3.65).

3.4.5 SOLUTION OF SYNTHETIC EQUATIONS IN THE FORM OF "DEFORMED" NORMAL WAVES

The synthetic approach offers, in principle, the accuracy $\sim (k l_{\mathrm{B}})^{-1}$, which can be attained if the initial condition for (3.68) is set through the iterative procedure as in the method of split rays (Section 3.3):

$$C_a = C_a^{(0)} + C_a^{(1)} + C_a^{(2)} + \cdots \quad (a = 1, 2),$$

$$C_2^{(0)} = 0, \quad \frac{dC_1^{(0)}}{ds_1} + p_{11}^1 C_1^{(0)} = 0,$$

$$C_1^{(0)}(s_{1\mathrm{in}}) \neq 0, \quad C_2^{(m)} = ik_0^{-1} |\beta_1| \left(1 - \frac{n_2}{n_1} \right)^{-1} \qquad (3.69)$$

$$\times \left(\frac{dC_2^{(m-1)}}{ds_1} + p_{21}^1 C_1^{(m-1)} + p_{22}^1 C_2^{(m-1)} \right),$$

$$\frac{dC_1^{(m)}}{ds_1} + p_{11}^1 C_1^{(m)} = -p_{12}^1 C_2^{(m)}, \quad C_1^{(m)}(s_{1\mathrm{in}}) = 0 \quad (m \geq 1).$$

In the region of substantially strong anisotropy ($\delta \sim 1$) the coefficient $C_1^{(0)}$ is proportional to $v_{1g}^{-1/2} \Phi_1$, where v_{1g} is the wave group velocity and the factor Φ_1 satisfies the conservation law (3.29).

For a weak anisotropy, one may safely set $|\beta_1| n_1 = 1$ in (3.69). Notice that the coefficients p_{ab}^1 in (3.68) differ somewhat from p_{ab} in (3.14):

$$p_{11}^1 = p_{11} + \tfrac{1}{2}\operatorname{div}\mathbf{t} + \tfrac{1}{2}d\ln n_1/ds, \quad p_{12}^1 = p_{12},$$

$$p_{22}^1 = p_{22} + \tfrac{1}{2}\operatorname{div}\mathbf{t} + \tfrac{1}{2}d\ln n_1/ds, \quad p_{21}^1 = p_{21}. \tag{3.70}$$

Additional terms in the coefficients p_{11}^1 and p_{12}^1 can easily be eliminated if the eikonal substitution (3.67) is replaced by

$$\mathcal{E} = \Phi_1 n_1^{-1/2}(C_1\mathbf{e}_1 + C_2\mathbf{e}_2)\exp{(i\varphi)}. \tag{3.71}$$

One can easily see that "synthetic" equations in a weakly anisotropic medium reduce to the analog of Budden's equations (3.12a) derived in this chapter from Eqs. (3.31) of the modified QIA method.

The reader may verify that in a plane-layered medium the "synthetic" equations reduce exactly to Budden's equations.

To conclude, we note that the variant of the QIA with split rays is a subasymptotics of the "synthetic" approach at $\mu_1 \ll \mu$. Although the original equations of the QIA are, strictly speaking, not valid at $\mu \sim 1$, nevertheless they may give qualitatively correct results for media with moderate anisotropy. This is favored by the availability of the iterative procedure with seeds, of the type of Eqs. (3.17)–(3.18), which raises the accuracy of the approximation.

Also, we note that the original version of the QIA, which does not incorporate ray splitting, nevertheless gives correct values for coefficients of transformation of normal waves. The account of ray splitting raises the accuracy of computations of phase difference between rays which undergone splitting and in this way allows one to describe more precisely the interference pattern of normal waves at an observation point.

4

Electromagnetic waves in an inhomogeneous plasma in a weak magnetic field

4.1 Quasi-longitudinal and quasi-transverse propagation

4.1.1 TENSOR OF PLASMA ANISOTROPY IN A WEAK MAGNETIC FIELD

Let us assume that the vector of an external magnetic field $\mathbf{H}^0$ is aligned with a coordinate plane (y, z). Then, the components of a Hermitian tensor $\widehat{\varepsilon}$ of a magnetoactive electron plasma take the form (Ginzburg, 1970):

$$\varepsilon_{xx} = 1 - \frac{v}{1-u}, \qquad \varepsilon_{yy} = 1 - \frac{v(1 - u\sin^2\alpha)}{1-u},$$

$$\varepsilon_{zz} = 1 - \frac{v(1 - u\cos^2\alpha)}{1-u}, \qquad \varepsilon_{xy} = -\varepsilon_{yx} = \frac{iv\sqrt{u}\cos\alpha}{1-u}, \qquad (4.1)$$

$$\varepsilon_{xz} = -\varepsilon_{zx} = -\frac{iv\sqrt{u}\sin\alpha}{1-u}, \qquad \varepsilon_{yz} = \varepsilon_{zy} = \frac{uv\cos\alpha\sin\alpha}{1-u},$$

where

$$u = \omega_{\mathrm{H}}^2/\omega^2 = (eH^0/mc\omega)^2, \quad v = \omega_0^2/\omega^2 = 4\pi e^2 N_e/m\omega^2, \qquad (4.2)$$

$\mathbf{H}^0$ is the vector of static magnetic field, N_e is the electron distribution density, e, m are the electron charge and mass, respectively; ω_{H} and ω_0 denote, respectively, the Larmor and plasma frequencies. α stands for the angle between the vectors $\mathbf{k}$ and $\mathbf{H}^0$. Its sign is defined as $\mathrm{sgn}\,\alpha = \mathrm{sgn}\,H_y$ if $|\alpha| \in [0, \pi)$.

The condition of weak anisotropy (1.3) in a weak magnetic field is satisfied for

$$u^{1/2} \ll 1 - v, \qquad v < 1. \tag{4.3}$$

In this case the components of tensor $\hat{\varepsilon}$ reduce to

$$\varepsilon_{xx} = (1 - v) - uv, \qquad \varepsilon_{yy} = (1 - v) - uv \cos^2 \alpha,$$

$$\varepsilon_{xy} = -\varepsilon_{yx} = iv\sqrt{u} \cos \alpha, \qquad \varepsilon_{xz} = -\varepsilon_{zx} = -iv\sqrt{u} \sin \alpha, \tag{4.4}$$

$$\varepsilon_{yz} = \varepsilon_{zy} = uv \cos \alpha \sin \alpha, \qquad \varepsilon_{zz} = (1 - v) - uv \sin^2 \alpha.$$

Terms of $3/2$ or higher powers in u in these expressions are discarded.

Another example of weak anisotropy is given by a rarefied plasma with parameters $v \ll 1, 1 - u \gg v$. We leave this case, which is encountered infrequently, as well as the case of a multicomponent plasma without consideration.

The components of the inverse tensor $\hat{\chi} = \hat{\varepsilon}^{-1}$ can be obtained from (4.1). We write here only those that will be further needed:

$$\chi_{xx} = (1 - v)^{-1} + (1 - v)^{-3} uv,$$

$$\chi_{yy} = (1 - v)^{-1} + (1 - v)^{-3} uv \cos^2 \alpha, \tag{4.5}$$

$$\chi_{xy} = -\chi_{yx} = -i(1 - v)^{-2} v\sqrt{u} \cos \alpha.$$

4.1.2   QIA EQUATIONS FOR A MAGNETOACTIVE PLASMA

To derive the QIA equations in the form (2.65), (2.66) we take the unit vectors $\mathbf{q}_1$ and $\mathbf{q}_2$ so that they form a right coordinate triple with the tangent unit vector $\boldsymbol{\tau}$, and the unit vector $\mathbf{q}_2$ lies in the plane $\boldsymbol{\tau}, \mathbf{H}^0$, tangent to both the ray and the static magnetic field vector $\mathbf{H}^0$ (see Figure 4.1). Then, replacing indices as $x \to 1, y \to 2$ in (4.4) and setting $\varepsilon_0 = 1 - v$ yields the following expressions for the coefficients of (2.66):

$$\gamma_{11} = -\gamma_{22} = -\frac{1}{2} uv \sin^2 \alpha, \quad \gamma_{12} = -\gamma_{21} = ivu^{1/2} \cos \alpha. \tag{4.6}$$

If we use the QIA equations for the electric induction vector, then the tensor $\tilde{\gamma}_{ik}$ given by (3.6) should be employed instead of the tensor γ_{ik}. Using (4.5) we replace (4.6) by:

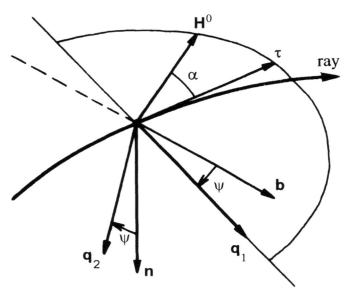

FIGURE 4.1. Mutual arrangement of the tangent to the ray τ, the normal n, the binormal b, the external magnetic field vector H^0, and auxiliary unit vectors q_1 and q_2.

$$\tilde{\gamma}_{11} = -\tilde{\gamma}_{22} = -\frac{1}{2}(1-v)^{-1}uv\sin^2\alpha\,,$$

$$\tilde{\gamma}_{12} = -\tilde{\gamma}_{21} = ivu^{1/2}\cos\alpha\,. \tag{4.7}$$

The component $\tilde{\gamma}_{11}$ in (4.7) differs from the corresponding component γ_{11} by the factor $(1-v)^{-1}$. This factor grows indefinitely as $v \to 1$.

Notice that, as $v \to 1$, the basic form of QIA equations (2.66) fails since the condition (2.70) breaks down. It is precisely to guarantee the applicability of results also for $v \sim 1$ that one needs the QIA equations for the induction vector D, i.e., the QIA equations in the form (3.7). Then

$$\mathcal{E} = \Phi_0\varepsilon_0^{-1/4}(\Gamma_1 q_1 + \Gamma_2 q_2)$$

$$\times \exp\left\{ik_0\int\left[(1-v)^{1/2} - \frac{uv(1+\cos^2\alpha)}{4(1-v)^{3/2}}\right]ds\right\}, \tag{4.8}$$

$$\frac{d\Gamma_1}{ds} = -\frac{1}{2} i\Gamma_1 G(1-v)^{-1} u^{1/2} \sin^2 \alpha - (G\cos\alpha - T_{\text{eff}}^{-1})\Gamma_2 \,,$$

$$\frac{d\Gamma_2}{ds} = (G\cos\alpha - T_{\text{eff}}^{-1})\Gamma_1 + \frac{1}{2} i\Gamma_2 G(1-v)^{-1} u^{1/2} \sin^2 \alpha \,. \tag{4.9}$$

Here $\mathcal{E} = \varepsilon_0^{-1}\mathcal{D}$ stands for the normalized electric induction vector, ψ is the angle between the binormal to the ray, $\mathbf{b}$, and the plane $\mathbf{t}, \mathbf{H}^0$ (shown in Figure 4.1); $T_{\text{eff}}^{-1} = T^{-1} + d\psi/ds$ is the effective torsion of the ray taking account of the rotation of magnetic field strength lines around the ray. The parameter G in Eqs. (4.9) denotes the combination

$$G = \frac{1}{2} k_0 (1-v)^{-1/2} v u^{1/2} \tag{4.10}$$

which characterizes the "strength" of field polarization variations in an inhomogeneous magnetoactive plasma.

Using Eqs. (4.9), we consider the mutual transformation of circular polarized waves on passing through what is called as the region of quasi-transverse propagation. Before proceeding with calculations we perform a preliminary qualitative analysis of the effect.

4.1.3   QUALITATIVE PICTURE OF THE INTERACTION OF CIRCULAR POLARIZED WAVES IN AN INHOMOGENEOUS MAGNETOACTIVE PLASMA

This picture, is in fact discussed in many works, beginning with the fundamental work by Budden (1952). Here we resort to the partial qualitative analysis performed by Zheleznyakov, Kocharovskii and Kocharovskii (1983).

The propagation of an electromagnetic wave through a homogeneous magnetoactive plasma is described most simply in two limiting cases:

(a) of the longitudinal propagation (the wave-vector $\mathbf{k}$ is parallel to the external magnetic field $\mathbf{H}^0$);

(b) of transverse propagation (the wave-vector $\mathbf{k}$ is perpendicular to $\mathbf{H}^0$).

For the longitudinal propagation, eigenwaves are circularly polarized, which corresponds to the Faraday effect. The transverse propagation is characterized by a linear polarization of eigenwaves (the Cotton–Mouton effect). In intermediate cases the waves are elliptically polarized.

In an inhomogeneous plasma, a transformation of polarization of normal wave occurs on both sides of the orthogonality point Q at which $\mathbf{H}^0 \perp \mathbf{k}$. A wave which is initially right circularly polarized (far from the point Q), changes to elliptically polarized, then (at the point Q) it becomes linearly polarized, then again it changes to an elliptically polarized wave (with the left direction of rotation), and, ultimately, to that left circularly polarized (see Figure 4.2). The evolution of the wave initially left circularly polarized occurs in the reverse order. Denote the characteristic length of the interval where the transformation of polarization occurs by $l_\perp$. The length $l_\perp$ coincides with the scale $l_p(\mathbf{r})$ introduced in Section 1.2.2 if the latter is taken on the ray at the point where $|\nu_{12}|$ is a minimum, and with the scale l_{int} introduced in Section 2.5.4. It follows from Eq. (2.7) that the difference of refractive indices, $\Delta n = |n_1 - n_2|$, reaches a local minimum (which is zero) at the point of transverse propagation where $\cos\alpha = 0$. The parameter $\delta = l_\perp^{-1} k_0^{-1} |n_1 - n_2|$ also has a local minimum there. Therefore, the most effective mutual transformation of ordinary and extraordinary waves occurs in the region just indicated.

One can easily find a parameter that defines the intensity of mutual transformation of right- and left-polarized waves in the interval of polarization transformation $l_\perp$. Its role plays the phase difference over this interval which we denote by p for the sake of convenience. If, for example, the parameter p is large,

$$p = k_0 l_\perp |n_1 - n_2|\Big|_Q = k_0 l_\perp uv(1-v)^{-3/2}\Big|_Q \gg 1\,,$$

then the condition (2.23) of normal wave independence is fulfilled everywhere in the interval of polarization transformation. Hence, normal waves almost do not transform at the orthogonality point Q. This implies that the ordinary wave, say such that was right-polarized far from the point Q, remains ordinary, but becomes left-polarized after passing through the point Q, while the extraordinary left-polarized one remains extraordinary, but becomes right-polarized. As a result, the coefficient of mutual transformation of right- and left-polarized waves is close to unity, while the coefficient of transformation of an ordinary wave into an extraordinary one is close to zero.

If the parameter p is small, $k_0 l_\perp |n_1 - n_2|\Big|_Q \ll 1$, the phase difference between the components with different polarization over the interval considered is close to zero. Therefore right- and left-polarized waves do not detect the interval of polarization transformation. In other words, circularly polarized waves preserve their polarization almost un-

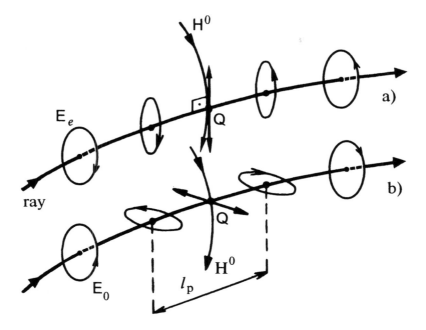

FIGURE 4.2. Evolution of polarization of normal waves in the vicinity of the orthogonality point Q. (a) An extraordinary wave which is initially right circularly polarized first converts into elliptically polarized wave, then it becomes linearly polarized, then again elliptically polarized and, finally, it converts to the right circularly polarized wave. (b) The evolution of an ordinary wave which is initially right circularly polarized occurs in inverse order.

changed, but change their "name": an ordinary wave converts into an extraordinary one, and vice versa. The coefficient of mutual transformation between ordinary and extraordinary waves is, accordingly, close to unity.

   The picture of interaction between polarization modes becomes markedly simpler in the case of the weak external magnetic field when the condition (4.3) holds. This case is commonly encountered in the solar atmosphere and the Earth's ionosphere. In a weak magnetic field the polarization of eigenwaves is close to circular over a wide range of angles α between the wave-vector $\mathbf{k}$ and the external field $\mathbf{H}^0$. The difference of the polarization from a circular one becomes noticeable only within the cone:

$$2|\cos\alpha| \lesssim u^{1/2}(1-v)^{-1}\sin^2\alpha. \qquad (4.11)$$

The interval where condition (4.11) is implemented is called the region of quasi-transverse wave propagation. In a weak magnetic field the dimension $l_\perp$ of this region is considerably smaller than the curvature radii of the ray or magnetic strength lines.

Indeed, in Eqs. (4.9), the terms $G\Gamma_{1,2}\cos\alpha$ correspond to the Faraday effect, while the terms $(1/2)G(1-v)^{-1}u^{1/2}\Gamma_{1,2}\sin^2\alpha$ are related to the Cotton–Mouton effect. "Faraday" terms prevail over their "Cotton–Mouton" counterparts for so-termed quasi-longitudinal propagation, when $2|\cos\alpha| \gg (1-v)^{-1}u^{1/2}\sin^2\alpha$. In this case the diagonal terms in Eqs. (4.9) are small and the equations describe an independent propagation of two circularly polarized waves. In the range of quasi-transverse propagation, where the inverse inequality (4.11) holds true, Cotton–Mouton diagonal terms, responsible for the wave transformation, are important in (4.3). On the other hand, by virtue of the assumption that the parameter $u^{1/2}(1-v)^{-1}$ is small, it follows from inequality (4.11) that the value of $\cos\alpha$ is also small everywhere in the quasi-transverse region. Based on this we may write

$$\cos\alpha = \frac{s}{\rho} + \mathcal{O}\left(\frac{s^3}{\rho^3}\right), \qquad \sin^2\alpha = 1 + \mathcal{O}\left(\frac{s^2}{\rho^2}\right), \qquad (4.12)$$

where s is the ray length measured from the orthogonality point $\cos\alpha = 0$, and $\rho = (d\cos\alpha/ds)^{-1}$ is the characteristic scale at which the angle α varies along the ray. It depends both on the ray curvature and on the magnetic field configuration. In particular, if the ray is completely in the plane of the magnetic meridian, $|\rho^{-1}| = |\rho_r^{-1} \pm \rho_H^{-1}|$, where ρ_r is the radius of ray curvature, and ρ_H is the curvature radius of magnetic lines, i.e., the distance from the orthogonality point to a virtual center of the magnetic field lines where the normals to the lines of $\mathbf{H}^0$ intersect.

It follows from expressions (4.12) and inequality (4.11) that in the interval of quasi-transverse propagation $2|s/\rho| \lesssim u^{1/2}(1-v)^{-1}$. Therefore the length of the interval of interaction satisfies the estimate

$$l_\perp \sim 2|s| \cong |\rho|u^{1/2}(1-v)^{-1} \ll |\rho|, \qquad (4.13)$$

i.e. $l_\perp$ is small compared to the curvature radius $|\rho|$.

We shall also assume that the inequality

$$l_\perp \ll l_0 = \min(l_H, l_N), \qquad (4.14)$$

is fulfilled, where l_0, l_H and l_N are the scales of inhomogeneities of the medium, magnetic field and plasma electron concentration, respectively. If inequality (4.14) holds true, one may speak of the spatial localization of the effect.

From estimate (4.13), the estimate for the parameter p that defines the intensity of transformation follows

$$p = k_0 l_\perp |n_1 - n_2|\Big|_Q \sim k_0 |\rho| u^{3/2} v(1 - v)^{-5/2}. \qquad (4.15)$$

For the case of $v \ll 1$ this estimate was first found by Cohen (1960). For the general case it was found by Melrose (1974).

4.2   Coefficients of transformation at the interval of quasi-transverse propagation

4.2.1   THE QIA EQUATIONS FOR WAVES WITH A CIRCULAR POLARIZATION

The effect of interaction of circularly polarized waves in a magnetoactive plasma in the region of quasi-transverse magnetic field (for brevity the effect will be referred to as "quasi-transverse" interaction) was invoked for the explanation of peculiarities of solar radiation in a radio band (Cohen, 1960; Zheleznyakov and Zlotnik, 1963; Zheleznyakov, 1970; Melrose, 1974) and some anomalies of the Faraday effect in the Earth's ionosphere (Kantor, Rie, Mendosa, 1970; Titcheridge, 1971). All calculations of this effect were conducted in the framework of a simplified statement of the problem which involved plane waves in a homogeneous plasma placed in an inhomogeneous magnetic field. The only significant analytical result for the transformation coefficients was found by Zheleznyakov and Zlotnik (1963) who used the phase integral method.

A more radical solution of the problem was suggested by Kravtsov and Naida (1976) and also by Naida (1978b).

Being interested in the transformation of circularly polarized waves, in Eq. (4.9) we change to the variables

$$\gamma_{1,2} = 2^{-1/2}(\Gamma_2 \mp i\Gamma_1). \qquad (4.16)$$

They obey the equations

$$\frac{d\gamma_1}{ds} = i(G\cos\alpha - T_{\text{eff}}^{-1})\gamma_1 + \frac{1}{2}iGu^{1/2}(1-v)^{-1}\gamma_2\sin^2\alpha\,,$$

$$\tag{4.17}$$

$$\frac{d\gamma_2}{ds} = \frac{1}{2}iGu^{1/2}(1-v)^{-1}\gamma_1\sin^2\alpha - i(G\cos\alpha - T_{\text{eff}}^{-1})\gamma_2\,,$$

and the normalization condition $|\gamma_1|^2 + |\gamma_2|^2 = 1$. If a wave incident from the side of negative s is right circularly polarized, the system of equations (4.10) should be supplemented by the initial conditions

$$|\gamma_1(-\infty)| = |\gamma_1^{\text{r}}(-\infty)| = 1\,, \qquad \gamma_2(-\infty) = \gamma_2^{\text{r}}(-\infty) = 0\,, \tag{4.18}$$

while for left circular polarization the initial conditions are

$$\gamma_1(-\infty) = \gamma_1^{\text{l}}(-\infty) = 0\,, \qquad |\gamma_2(-\infty)| = |\gamma_2^{\text{l}}(-\infty)| = 1\,. \tag{4.19}$$

4.2.2   ANALYTICAL SOLUTION OF THE QIA EQUATIONS FOR LOCALIZED INTERACTION

The system of equations (4.17) is a particular case of the system (2.71) and does not yield to analytical solution for arbitrary variable parameters. However, if the region of interaction is localized, i.e. the length of the interaction region, $l_\perp$, is small compared not only to ρ, but also to a characteristic scale of plasma inhomogeneity, l_0, the system (4.17) admits an approximate solution, possessing, nevertheless, a reasonable universality. In fact, for $l_\perp \ll l_0$ the plasma parameters u and v, as well as the effective torsion T_{eff}^{-1} within the region of interaction can be replaced by their local values at the point of orthogonality. Then, in conjunction with (4.12), we find the system of equations

$$\frac{d\gamma_1}{ds} = i(G\rho^{-1}s - T_{\text{eff}}^{-1})\gamma_1 + \frac{1}{2}iG(1-v)^{-1}u^{1/2}\gamma_2\,,$$

$$\tag{4.20}$$

$$\frac{d\gamma_2}{ds} = \frac{1}{2}iG(1-v)^{-1}u^{1/2}\gamma_1 - i(G\rho^{-1}s - T_{\text{eff}}^{-1})\gamma_2\,.$$

where according to (4.10)

$$G = \frac{1}{2}k_0 v u^{1/2}(1-v)^{-1/2}\,.$$

If a dimensionless variable

$$\xi = \left(\frac{G}{|\rho|}\right)^{1/2} (s - \rho G^{-1} T_{\text{eff}}^{-1}), \tag{4.21}$$

is introduced, Eqs. (4.20) take the form

$$\frac{d\gamma_1}{d\xi} = i\xi\gamma_1 \operatorname{sgn}\rho + \frac{1}{2} i p^{1/2}\gamma_2 ,$$

$$\frac{d\gamma_2}{d\xi} = \frac{1}{2} i p^{1/2}\gamma_1 - i\xi\gamma_2 \operatorname{sgn}\rho , \tag{4.22}$$

i.e. the form of the type of the system (2.85). Eqs. (4.22) contain a single parameter,

$$p = Gu|\rho|(1-v)^{-2} = \frac{1}{2} k_0 v u^{3/2}|\rho|(1-v)^{-5/2} . \tag{4.23}$$

As one would expect, formula (4.23) agrees with the estimate (4.15) whereas Eqs. (4.20) and the concomitant formula (4.21) form a particular case of Eqs. (2.80) and its "satellite" (2.83). Correspondingly, Eq. (4.23) is in complete agreement with Eq. (2.86).

Accordingly, the component γ_2 of system (4.22) satisfies the Weber–Hermite equation and the solution of the system of equations (4.22) with initial conditions (4.18) (initial right circular polarization) can be expressed in terms of functions of the parabolic cylinder $D_n(z)$, whereas the intensities of circularly polarized waves for $\xi \to +\infty$ are equal to

$$\left|\gamma_1^r(+\infty)\right|^2 = \exp\left(-\frac{\pi p}{4}\right) ,$$

$$\left|\gamma_2^l(+\infty)\right|^2 = 1 - \exp\left(-\frac{\pi p}{4}\right) . \tag{4.24}$$

Hence, the quantity

$$\eta = 1 - \exp\left(-\frac{\pi p}{4}\right) \tag{4.25}$$

represents the coefficient of mutual transformation of a right-polarized wave into the left-polarized wave and vice versa.

Formula (4.25) coincides with an expression for the coefficient of transformation obtained in the framework of the solution of a 1D problem by the method of phase integrals (Zheleznyakov and Zlotnik, 1963, 1977), although the results of the QIA are in many respects more comprehensive. First, the QIA applies to 3D curved rays. Second, the QIA not only gives the value of the transformation coefficient η, but also the magnitude of the field at all points of the ray. That makes it possible to calculate η even in the cases of a source located inside the region of interaction. This has important implications for ionospheric investigations. Third, it turns out that torsion of the ray and the rotation of the magnetic field strength lines do not influence the value of η, since the effective torsion, T_{eff}^{-1}, has dropped out from Eq. (4.22). Finally we may derive that the wave transformation is small in an extended region of quasi-longitudinal propagation and find corrections to transformation coefficient (4.25) due to "partial localization" of the interaction effect from the basic equations (4.17). We return to this question later.

4.2.3   POSSIBILITIES OF "COTTON–MOUTON" PLASMA DIAGNOSTICS

Unlike the Faraday effect which provides information about *integrated* parameters of the plasma, the quasi-transverse interaction due to the Cotton–Mouton effect may serve as a source of information about *local* characteristics of plasma at the orthogonality point. In particular, one may speak about determination of the local electron concentration N_e with the other parameters being known.

If the effect of quasi-transverse interaction is spatially localized (condition (4.14)), the coefficient of transformation is given by (4.25), and one may determine the parameter p which characterizes local plasma properties at the point of orthogonality by measured values of η:

$$p \equiv G_\circ u_\circ |\rho| = \frac{k v_\circ u_\circ^{3/2} |\rho|}{2} = \frac{4}{\pi} \ln \frac{1}{1-\eta}, \quad (v \ll 1). \qquad (4.26)$$

The possibility of that kind, substantiated by the results of Cohen (1960), and Zheleznyakov and Zlotnik (1963), is widely and successfully used to treat the data on the polarization of solar irradiance, in particular, to estimate the magnetic field H_0 of the solar corona based on preliminary estimates of other plasma parameters. However, as applied to ionospheric and laboratory plasmas, the possibilities of Cotton–Mouton diagnostics have not been discussed in detail, particularly, due to the lack of an effective theory for normal wave transformation in the case of curved rays.

Under the conditions of the Earth's ionosphere the characteristic scale ρ is 3000–6000 km (given a dipole model of the magnetic field), whereas the vertical scale for plasma inhomogeneities $l_{vert} \sim 100$ km. By virtue of (4.13) the effect of quasi-transverse interaction will be localized for $u^{1/2} \ll l_{\perp}/|\rho| \sim 1/30 - 1/60$, i.e. for $\lambda \lesssim 4 - 7$ m. Thus, in the ultrahigh frequency range, one can rely on measurements of local electron concentration of an ionospheric plasma at the points of orthogonality. For an oblique propagation of radiowaves the effective scale l_0 of inhomogeneities increases several times (since $l_{horiz} \sim 1000$ km), and the threshold wavelength also increases in the same manner (up to 20–30 m).

For a laboratory plasma, the values N_e, H_0, and ρ may vary within extremely wide limits; hence it seems quite reasonable that with an appropriate choice of frequency and propagation direction for microwave radiation large volumes of plasma can be investigated with a view to determine the local electron concentration. That could be done in a favourable way by intentionally controlling the magnitude and configuration of the magnetic field provided it is admissible under the conditions of experiment.

In solving problems of plasma diagnosis, one should also, in addition to the localization of the effect (condition (4.14)), strive for its better "visibility" which could be, for instance, characterized by the degree of linear polarization $\beta_{lin} = 2[\eta(1-\eta)]^{1/2}$. This value reaches a maximum at $\eta = 1/2$ which is attained for $p_{1/2} = 4\pi^{-1} \ln 2 = 0.88$ and for the frequency of radiation ω_1

$$\frac{\omega_1}{\omega_0} = (0.57\omega_0^{-2}\omega_H^3 |\rho|c^{-1})^{1/4}.$$

Generally speaking, inequalities (4.3) or (4.14) can be violated at this frequency which offers "best visibility" of the effect. In order to avoid such a violation it is sufficient for two conditions $\omega_0/\omega_1 \ll 1$ and $\omega_H/\omega_1 \ll 1$ to be satisfied simultaneously. This is equivalent to the following double constraint on the plasma frequency ω_0

$$\omega_H^{3/2} \left(\frac{|\rho|}{c} \right)^{1/2} \gg \omega_0 \gg \omega_H^{1/2} \left(\frac{|\rho|}{c} \right)^{-1/2},$$

which, in turn, is possible if $\omega_H |\rho|/c \gg 1$. These inequalities specify the ranges of values of ω_H and ω_0, for which both the spatial localization and good visibility of the effect are possible.

4.2.4 ACCOUNTING FOR THE PARTIAL LOCALIZATION OF THE COTTON–MOUTON EFFECT

Estimates of errors due to the replacement of the exact QIA equations (4.17) by Eqs. (4.20) with linearized coefficients can be calculated in various ways. In essence, to make these estimates we should take into account the influence of nonlinear in s terms in the coefficients of Eq. (4.17). Denoting the maximum absolute value of these terms as δP, we can write:

$$\delta P \sim G \max \left\{ \frac{s^3}{|\rho|^3}, \sqrt{u}\frac{s^2}{\rho^2}, \frac{s^2}{l_0|\rho|} \right\}.$$

The first term in the brackets corresponds to a cubic term omitted when expanding $\cos \alpha$ in s/ρ, the second term derives from the expansion of $\sin^2 \alpha$ in s/ρ and the third one corresponds to a linear in s term in the expansion of the coefficient G (it is then multiplied by $\cos \alpha \sim s/\rho$).

The simplest way to evaluate errors $\delta\gamma$ introduced by the corrections δP in the solution γ and, correspondingly, in the transformation coefficient η ($\delta\eta \sim \delta\gamma$) is to separately evaluate the error $\delta\gamma_1$ associated with the region of weak interaction of normal waves and the error $\delta\gamma_2$ associated with the region of moderate and strong interaction, and then choose the largest.

We restrict ourselves to the simplest case when $1-v \sim 1$ and the visibility is ~ 1, i.e. $p \sim 1$. By virtue of the latter assumption the boundary between the regions of quasi-longitudinal and quasi-transverse propagation coincides with the boundary of convergence of the "seed" iteration (3.17)–(3.19). The first of formulae (3.19) gives the estimate of $\delta\gamma_1$:

$$\delta\gamma_1 = \left[\frac{\delta P}{l_0 \Delta n_0} \right]_{\mathrm{B}} \sim \frac{k_0 v u^{1/2}}{k_0 u v} \max \left(u^{3/2}, u\frac{|\rho|}{l_0} \right) = \max \left(u, \frac{|\rho|}{l_0}\sqrt{u} \right).$$

The symbol B here denotes the boundary between the regions of quasi-longitudinal and quasi-transverse propagation.

Now we obtain

$$\delta\gamma_2 \sim (\delta P)_{\mathrm{B}} l_\perp \sim \delta\gamma_1,$$

since in the case considered $k_0(\Delta n_0)_{\mathrm{B}} l_\perp \sim 1$. Thus

$$\delta\eta \sim \max \left(u, \frac{|\rho|}{l_0}\sqrt{u} \right).$$

4.2.5   Ionospheric Manifestations of the Cotton–Mouton Effect. Weak Depolarization of Radiowaves

Already the first works devoted to the QIA were intended to give a description of radiowave polarization in the ionosphere. A simple way to describe a weak depolarization caused by a distributed, non-localized Cotton–Mouton effect was suggested in a work by Kravtsov (1968a).

Let us write the Riccati equation (2.51) for a plasma placed in a weak magnetic field. In compliance with (4.1), we find the equation for ϑ:

$$\frac{d\vartheta}{ds} = -T^{-1} + \frac{1}{2}k_0 v(1-v)^{-1}u^{1/2}\cos\alpha$$

$$+ \frac{i}{4}k_0 vu\sin^2\alpha\sin 2(\vartheta + \psi)\,. \tag{4.27}$$

The first term in that equation describes the Rytov torsion, the second one describes the Faraday rotation of the polarization plane, whereas the third imaginary term corresponds to the Cotton–Mouton effect.

For most rays the third term is small compared to the second. Hence an iterative procedure yields a solution to Eq. (4.27). For that purpose we write Eq. (4.27) in the form

$$\frac{d\vartheta}{ds} = -T^{-1} + \frac{1}{2}k_0 v(1-v)^{-1}u^{1/2}\cos\alpha + iM(s,\vartheta)\,, \tag{4.28}$$

where $M(s,\vartheta) = -(1/4)k_0 vu\sin^2\alpha\sin 2(\vartheta + \psi)$. Setting $M = 0$, we obtain the zeroth approximation

$$\vartheta_0(s) = \vartheta(0) - \int_0^s \frac{ds}{T} + \vartheta_{\mathrm{F}}(s)\,, \tag{4.29}$$

where ϑ_{F} is the Faraday rotation angle,

$$\vartheta_{\mathrm{F}}(s) = \frac{1}{2}k_0 \int_0^s v(1-v)^{-1}u^{1/2}\cos\alpha\,ds\,. \tag{4.30}$$

The first iteration already results in a small imaginary correction to $\vartheta_0(s)$:

$$\vartheta_1(s) = \vartheta_0(s) + i\int_0^s M(s,\vartheta_0)\,ds$$

$$= \vartheta_0(s) - \frac{i}{4}k_0 \int_0^s vu\sin^2\alpha\sin 2(\vartheta_0 + \psi)\,ds\,. \tag{4.31}$$

The imaginary correction $\vartheta_1(s) = i\vartheta_1''$ is responsible for the transformation of a linearly polarized wave into an elliptically polarized wave, with the small axis

$$\tan \vartheta_1'' \approx \text{Im } \vartheta_1 = -\frac{1}{4} k_0 \int_0^s vu \sin^2 \alpha \sin 2(\vartheta_0 + \psi) \, ds \, . \qquad (4.32)$$

Thus, for the ionospheric propagation of radiowaves the state of polarization is characterized by the angle (4.29) (which is the sum of the Rytov and Faraday angles) through which the large axis of polarization ellipse turns and by the small axis (4.32) that defines the depolarization of the wave.

Fuki (1987a) has constructed another variant of the iterative procedure which possesses an improved convergence. We rewrite Eq. (4.28) in the form of two coupled equations for $\vartheta' = \text{Re } \vartheta$ and $\vartheta'' = \text{Im } \vartheta$:

$$\frac{d\vartheta'}{ds} = -T^{-1} + \frac{d\vartheta_{\text{F}}}{ds} + \frac{1}{4} k_0 vu \sin^2 \alpha \cos 2(\vartheta' + \psi) \sinh 2\vartheta'' \, ,$$

$$\frac{d\vartheta''}{ds} = -\frac{1}{4} vu \sin^2 \alpha \sin 2(\vartheta' + \psi) \cosh \vartheta'' \, . \qquad (4.33)$$

Here ϑ_{F} is the Faraday rotation angle (4.30). From the second of the equations (4.33) it follows that

$$\sinh 2\vartheta'' = \tan\left[\arctan \sinh 2\vartheta''(0) - L(s, \vartheta')\right] \, , \qquad (4.34)$$

where

$$L(s, \vartheta') = \frac{1}{2} k_0 \int_0^s vu \sin^2 \alpha \sin 2(\vartheta' + \psi) \, ds \, ,$$

and $\vartheta''(0)$ is the initial value of ϑ''. In contract with the procedure just described above, where we set $\vartheta''(0) = 0$, the value of $\vartheta''(0)$ can be arbitrary.

On substituting (4.34) in the first of equations (4.33) we find a closed equation for ϑ' which can be rewritten in the integral form:

$$\vartheta'(s) = \vartheta'(0) - \int_0^s \frac{ds}{T} + \vartheta_{\text{F}}(s) + R(s, \vartheta') \, , \qquad (4.35)$$

where

$$R(s, \vartheta') = -\frac{1}{4} k_0 \int_0^s vu \sin^2 \alpha \cos 2(\vartheta' + \psi)$$

$$\times \tan\left[\arctan \sinh 2\vartheta''(0) - L(s, \vartheta')\right] ds \, .$$

Since R is small, it is natural to solve (4.35) by iterations. If, similarly to (4.29), we set

$$\vartheta_0'(s) = \vartheta'(0) - \int_0^s \frac{ds}{T} + \vartheta_F(s) \tag{4.36}$$

in the zeroth approximation then, in the m-th approximation

$$\vartheta_m'(s) = \vartheta_0'(s) + R(s, \vartheta_{m-1}'),$$

$$\sinh 2\vartheta_m''(s) = \tan\left[\arctan \sinh 2\vartheta''(0) - L(s, \vartheta_m')\right], \tag{4.37}$$

The process (4.37) converges rapidly, and it is sufficient to perform one or two iterations in most cases of practical importance (see Table 4.1 below).

In the simplest case, when an incident wave is linearly polarized and $\vartheta''(0) = 0$, Fuki's iterative procedure leads in the first approximation to

$$\vartheta_1'(s) \approx \theta_0' - \int_0^s \frac{ds}{T} + \vartheta_F(s) + \Delta_1\theta', \tag{4.37a}$$

where $\Delta_1\theta'$ is the absolute error of the first approximation

$$\Delta_1\theta' = |\theta' - \theta_1'| \lesssim \frac{1}{2}k_0 L_p (uv)_{\max}; \tag{4.37b}$$

here L_p is the path length and the index "max" denotes maximum magnitudes of the parameters over the path.

It is clear that the solution (4.37a) coincides with the solution (4.31), and (4.37b) gives the estimate of the inaccuracy of the latter. We note that this inaccuracy declines rapidly as the frequency grows ($\sim \omega^6$).

In the second approximation

$$\vartheta_2' \approx \theta_1' - \frac{k_0}{4} \int_0^s uv \cos 2(\theta_1' + \psi) \tan L(s, \theta') + \Delta_2\theta' \tag{4.37c},$$

where

$$\Delta_2\theta' = |\vartheta' - \vartheta_2'| \lesssim \frac{1}{32}(k_0 L_p uv)_{\max}^4. \tag{4.37d}$$

The magnitude of the error $\Delta_2\theta'$ decays even more rapidly when the frequency of the sounding signal grows ($\sim \omega^{12}$).

Fuki's iterative procedure in the first approximation gives the following expression for the magnitude of depolarization of the signal over the path (the change of the degree of ellipticity of the signal)

$$\tanh \theta'' \approx \tanh \theta_1'' = -\int_0^s uv \sin 2(\theta' + \psi)\, ds + \Delta_1\theta'', \qquad (4.37e)$$

where

$$\Delta_1\theta'' = |\tanh \theta'' - \tanh \theta_1''| \lesssim \frac{4}{13}(k_0 L_p uv)^3. \qquad (4.37f)$$

For many problems concerning the ionospheric propagation of high frequency and ultrahigh frequency radiowaves the approximation (4.32) offers a reasonably satisfactory calculation of the depolarization. Otherwise (if the parameter M in (4.28) is large enough) the iterative scheme (4.37) should be used.

Tokar, Rubinshtein and Nikitin (1987) performed detailed calculations of radiowave polarization for specific models of the ionosphere. The QIA equations were solved numerically by the Runge–Kutta method. According to this work, the Rytov rotation of the polarization plane can be disregarded in most cases compared to the Faraday rotation, since the radius of the ray torsion, T, is frequently extremely large even under the conditions of transition from the diurnal region of the ionosphere to the nocturnal region. It is then that the contribution from the torsion is maximum, for the structure of diurnal and nocturnal ionosphere is plane-layered, and thus torsion is practically absent.

4.2.6   POLARIZATION EFFECTS DUE TO WAVE SCATTERING IN HIGH-LATITUDE AND EQUATORIAL IONOSPHERE

One often needs to account for changes in the polarization of radio waves scattered on irregularities of the high-latitude and equatorial ionosphere. As shown by Goryshnik and Kravtsov (1969) these changes stem from two factors: (i) the transformation of waves as they propagate from the source and return (integral mechanism of scattering) and (ii) the anisotropy of medium·irregularities just at the point of scattering (local mechanism of scattering). The local depolarization is usually negligible at the point of scattering and therefore the resultant depolarization is due to the integral mechanism, i.e. to the polarization changes along the entire ray.

Later on, Fuki (1987a, b, 1988) refined the results obtained by Goryshnik and Kravtsov (1969) and discovered another bulk mechanism of depolarization characteristic of signals with long duration. It is known that for a radar the auroral radioecho (radioaurora) represents a volume-distributed target consisting of many elements. These elements originate from small-scale anisotropic irregularities of the ionospheric plasma, on which the signal is in fact scattered (i. e. auroral irregularities). If the pulse volume is large, the plane of polarization of the electromagnetic wave may turn over a noticeable Faraday angle. Therefore signals scattered on irregularities at the front and back parts of the pulse differ not only in phase, but also in their polarization. This leads to bulk depolarization. Its calculations can be reduced to averaging the polarization characteristics of the signal over the pulse volume. The bulk mechanism of depolarization proves to be insignificant for very short pulses, with the length much less than the period of Faraday rotation of the polarization plane.

Fuki (1987a, b, 1988) obtained the correlation matrix of scattered field $< E_{sj} E_{sj}^* >$ taking into account all principal effects which influence the scattering of signals in high latitudes, and on the basis of this solved the problem of depolarization (change in the degree of ellipticity) of signals scattered in the high-latitude atmosphere. The analysis complies with the known fact that the main reason for the polarization change is the Faraday effect. Specifically, it allows one to estimate the electron concentration in the ionosphere by the rotation of the polarization plane of the signal scattered on auroral irregularities.

According to Fuki (1987a, b, 1988) sufficiently high magnitudes of the depolarization of the signal detected in experiments performed by Goryshnik, Kravtsov, Tomashchuk and Fomin (1969) are most probably due to the bulk mechanism of depolarization whose relative contribution can be 30%. As the duration of the signal increases, the bulk mechanism becomes prevailing.

The effect of bulk depolarization may lead to an error in determining the Faraday rotation angle θ_F. Estimates performed for various profiles of electronic concentration in E-layer have shown that the relative error in determining θ_F can be as high as 10–35%.

When analyzing the polarization structure of the signal, Goryshnik, Kravtsov, Tomashchuk and Fomin (1969) actually employed the first-order approximation of the iterative procedure described in Section 4.2.5 (formulae (4.31), (4.32)). Table 4.1 lists the uncertainties in the first, $\Delta_1 \theta'$, and in the second, $\Delta_2 \theta'$, approximations (formulae (4.37b), (4.37d)) at frequencies of 58.3 and 74.1 MHz (used in the experiments

TABLE 4.1. Estimates of errors of the first and second approximations of the iterative procedure

f, MHz	$\Delta_1\theta'$, degree	$\Delta_2\theta'$, degree
58.3	4.2	0.8
74.1	1.0	$3.5\cdot10^{-2}$
150	$3.5\cdot10^{-2}$	$6\cdot10^{-4}$

mentioned) and also at a frequency of 150 MHz (frequently used in radioaurora investigations).

Table 4.1 shows that errors in determining $\Delta_1\theta'$ in the first approximation of the iterative procedure amount to hundredths of one degree so one does not need the second approximation. In the same time, the inaccuracy of the first approximation makes up $4°$ at a frequency of 58.3 MHz, a value noticeably larger than the experimental error (about $0.5°$). The error of the second approximation of $0.6°$ is still comparable with the experimental error, which implies that one should resort to the third iteration in this case.

In the case when the radar signal scatters on small-scale irregularities of the auroral ionosphere, the polarization angle θ becomes a random quantity, thus hampering the assessment of the electron concentration by polarization characteristics of the signal. Statistical methods which allow one to optimize estimates of θ can be conveniently used under these conditions (Fuki 1987c, 1988; Kravtsov and Fuki, 1990).

In addition to electron concentration, statistical methods also give an estimation of the thickness of the scattering region (see Fuki (1988, 1990a, b)). To show this we consider the expressions for the variances of orthogonal field components $\sigma_x^2 =< |\mathbf{E}_x|^2 >$ and $\sigma_y^2 =< |\mathbf{E}_y|^2 >$ (Fuki, 1988):

$$\begin{pmatrix} \sigma_x^2 \\ \sigma_y^2 \end{pmatrix} = A_1 \int_V dV \frac{|g(\mathbf{n}-\mathbf{n}_0)|^2}{|\mathbf{r}-\mathbf{R}|^4} \left\{ \begin{array}{c} |\cos 2\theta|^2 \\ |\sin 2\theta|^2 \end{array} \right\} \Phi_N(\mathbf{q},\mathbf{R})|\mathcal{S}(t-2\Delta t)^2|,$$

(4.38)

and the correlation coefficient $\mathcal{R} =< E_x E_y^* >$ is

$$\mathcal{R} = \frac{A_1}{\sigma_x\sigma_y} \int_V dV \frac{|g(\mathbf{n}-\mathbf{n}_0)|^2}{|\mathbf{r}-\mathbf{R}|^4} \cos 2\theta \sin 2\theta^* \Phi_N(\mathbf{q},\mathbf{R})|\mathcal{S}(t-2\Delta t)^2|.$$

(4.39)

Here $A_1 = (\pi k_0^4/2)(\partial\varepsilon/\partial N) = 8\pi^3 e^4/m^2 c^4 \approx 1.97 \cdot 10^{-23}$ cm^2 (we use the Gaussian unit system), $\Delta t(\mathbf{R}) = |\mathbf{r}-\mathbf{R}|/c$ is the group delay, $\Phi_N(\mathbf{q},\mathbf{R})$ is the local spectral density of inhomogeneities in electron

concentration N_e, $\mathbf{q} = \mathbf{q(R)} = -2\mathbf{k}_0$ is the local scattering vector corresponding to the point $\mathbf{R}$ of the pulse volume V, $g(\mathbf{n} - \mathbf{n}_0)$ is the
directivity pattern of the antennae, $\mathbf{n}_0$ is the direction of its main lobe,
$\mathcal{S}(t)$ is the signal envelope. The turn of the field vector is adequately
described by (4.30). Fuki (1988) related the statistical characteristics
of the polarization of the received signal to the angles of rotation of
the field vector on the path to the scattering region (angle θ_i) and exactly within the pulse volume (angle $\Delta\theta$). Provided that the scattering
volume and angles θ_i and $\Delta\theta$ are all small, the mean magnitude of the
turn angle θ is

$$< \theta > = \theta_i + \frac{1}{2}\Delta\theta,$$

while the variance σ_θ^2 to the first order in $\Delta\theta$ is

$$\sigma_\theta^2 \approx \frac{\ln 2}{\sqrt{3}}\Delta\theta \left(1 + \theta_i^2 + \theta_i\Delta\theta - \frac{\sqrt{3}}{12\ln 2}\Delta\theta + \mathcal{O}(\Delta\theta^2) \right).$$

Knowing the experimental values of the variance σ_θ^2 and mean $< \theta >$,
one may determine $\Delta\theta$ and θ_i and then the distribution of the electron
concentration in the auroral backscattering E-layer and on the path to
the layer in the E-region, as well as other parameters of the radioaurora
(Fuki, 1988, 1990a, b).

The results of the correlation theory can also be used for the analysis of irregularities in the equatorial E-region, and with some modifications, can also be used to analyse incoherent (Thompson) scattering.
In the latter case one should employ Φ_N for the spectrum of thermal
fluctuations in the ionospheric plasma.

4.2.7 POLARIZATION METHOD OF AURORAL RADIOECHO REDUCTION

Strong backscattering of waves caused by anisotropic magnetic-oriented
irregularities in the electronic concentration of the E-region of the ionosphere presents a hindrance to radio engineering systems of different
types: wireless communication, radio location, radio navigation etc.
Kharitonova, Fuki and Bukatov (1989) suggested a polarization method
to reduce the influence of auroral radioecho.

The method stems from the fact that for a quasi-transverse (and
the more so, for quasi-longitudinal) propagation the depolarization of
circularly polarized waves is relatively small, i.e. the energy pumped
into the linear polarization is negligible. It is for this reason that the

influence of hindering radioecho turns out to be strongly reduced, while the signal does not suffer any such reduction.

The depolarization of a radiated circularly polarized wave is largely due to the integral mechanism. Indeed, for a circular polarization the resultant signal will be a sum of elementary signals with circular polarization. Since the sum of similar circular polarizations (with the same rotation sense) gives birth again to circular polarization; the resultant signal is circularly polarized, i.e. the bulk mechanism of depolarization does not alter the compression coefficient ($\tanh\theta''$) of a circularly polarized wave. Local depolarization, as shown by the authors mentioned, is also negligible. Therefore, a change in polarization is due to the integral mechanism and by virtue of (4.34) is described by

$$\tanh\theta''(s) = 1 - L + (1/2)L^2 - (1/3)L^3 + \mathcal{O}(L^4), \qquad (4.40)$$

where

$$L = \frac{1}{2}k_0 \int_0^s uv\sin^2\alpha\sin2(\theta' + \psi)ds. \qquad (4.41)$$

The functional (4.41) describes the Cotton–Mouton depolarization of the signal along its path. Its value decreases rapidly with growth of the frequency ($\sim \omega^{-3}$).

According to (4.40), a primary circularly polarized wave transforms into an elliptically polarized wave with the axis ratio $\rho = |\tanh\theta''|$. Such a transformation can be interpreted as a partial pumping of energy of the primary left-polarized wave into a right-polarized wave.

The transformation coefficient η_t describing this energy pumping is (Fuki, 1990):

$$\eta_t = \frac{1}{4}L^2(1 + \frac{1}{6}L^2 + \mathcal{O}(L^3)).$$

In the real ionosphere, the transformation coefficient is relatively small for waves with a very high frequency range: $\eta_{max} < -(35-45)$ dB. This substantiates the use of the orthogonal polarization selection method for suppressing the interference from auroral radioecho signals.

In practice, the quality of auroral radioecho suppression is related to the degree to which the polarization of the emitted signal is close to a circular one and depends on the decoupling between the orthogonal polarization channels in the receiver. The efficiency of the method can be characterized in the following way. The degree of auroral radioecho suppression in the receiver can reach -27dB for the main lobe width of the order of $1°$, a relative band of the signal emitted $\Delta f/f = 10\%$, provided the two orthogonal polarizations are decoupled completely in the receiver. If $\Delta f/f = 5\%$ it can reach -33 dB. When the decoupling between the receiving channels is 30 dB, the degree of suppression becomes -25 and -28 dB for the two values of the relative band respectively.

4.3 Other polarization effects in plasmas

4.3.1 THE TRANSFORMATION OF NORMAL WAVES IN THE REGION OF ZERO MAGNETIC FIELD IN THE SOLAR CORONA

We have confined ourselves above to considering only one, though important, question about changes in polarization in the region of quasi-transverse propagation of electromagnetic waves in a magnetoactive plasma. When analyzing the effects that could influence the polarization of electromagnetic waves radiated by the solar corona, Zheleznyakov (1964) discovered that a marked transformation of normal waves also occurs in regions of a neutral magnetic field. Such regions could be formed, say, by two or several current sources creating magnetic fields in different directions. In magnetic fields of complex structure, the presence of lines where the field $\mathbf{H}^0$ equals zero ("null-lines") is the rule rather than the exception.

A moving solar plasma moves a frozen magnetic field into a near-solar space, stretching the patches with neutral magnetic field $\mathbf{H}_0 = 0$ there as well.

In the vicinity of the neutral field point everything looks as if we have encountered the problem of limit polarization twice: first the normal waves propagating in the magnetoactive plasma become transverse as $\mathbf{H}^0 \to 0$, then a transverse wave splits into the superposition of two normal waves past the neutral point. The results of rigorous analysis of this problem are presented in a review article by Zheleznyakov, Kocharovskii and Kocharovskii (1983).

4.3.2 POLARIZATION EFFECTS IN A MOVING PLASMA

Inhomogeneous motion either of a laboratory or any other plasma results in a weak anisotropy. In essence, this anisotropy is similar to the optical Maxwell effect (appearance of anisotropy in the shear flow of a fluid). Mechanisms which give rise to anisotropy in a plasma are discussed in works by Stepanov and Gavrilenko (1971), Gavrilenko,

Lupanov and Stepanov (1972), Gavrilenko and Stepanov (1976). In essence, they stem from a spatial dispersion induced by an inhomogeneous flow of plasma. Clearly, polarization effects in an inhomogeneously moving plasma can be described in the framework of the QIA.

4.3.3   WEAK ANISOTROPY DUE TO THE INHOMOGENEITY OF THE MEDIUM

The inhomogeneity of a medium which possesses a spatial dispersion must inevitably lead to anisotropy, since there exists a distinguished direction $\nabla \varepsilon_0$, where ε_0 is the permittivity of the isotropic medium in the absence of inhomogeneities. Calculation of the plasma inhomogeneity tensor carried out by Kravtsov, Kugushev and Chernykh (1970) has confirmed this expectation. The anisotropy induced by inhomogeneity turned out to be negligible. However at large distances such anisotropy could be discovered by polarization methods. In essence, even the Rytov rotation of the polarization plane might be interpreted as being caused by the appearance of a weak anisotropy due to inhomogeneities of the medium (see also Section 5.4).

4.3.4   POLARIZATION EFFECTS IN A PLASMA WITH RANDOM INHOMOGENEITIES

In a work by Apresyan (1976) the problem of the influence of weak fluctuations in an anisotropic medium on the polarization of a wave passing through a thick randomly inhomogeneous layer has been considered. This problem is encountered, in particular, in describing the polarization of high-frequency waves in a magnetoactive cosmic plasma. In the work mentioned the QIA equations with a fluctuating tensor of dielectric permittivity were used. On this basis, equations for the mean value of the Stokes vector, which describes the wave polarization, were derived. As it turns out, the mean Stokes vector within the layer tends asymptotically to a specific direction, related to the mean value of the tensor of the dielectric permittivity of the medium. Thus the measurement of the mean polarization provides additional information about mean properties of the medium and in that way facilitates the solution of inverse problems. This problem is also briefly addressed in a book by Apresyan and Kravtsov (1996) and in a review by Apresyan and Kravtsov (1997).

4.3.5   INTERACTIONS OF ELECTROMAGNETIC WAVES NEAR CAUSTICS

It could be anticipated that the picture of mutual transformations of electromagnetic waves in an inhomogeneous plasma would undergo appreciable changes if a caustic is present in the interaction region, as frequently occurs in problems involving ionospheric wave propagation. The appearance of caustics does indeed have an effect on linear wave transformation, but, in many cases, the alterations in output characteristics are so insignificant that they can be ignored. One such case will be considered below.

The case in point is the wave transformation under conditions when ray splitting is negligible and one deals with the QIA equations in their standard form (2.48), with the arc length measured along the common "quasi-isotropic" ray.

A caustic appearing in the path of a family of such rays radically alters the structure of electromagnetic waves as compared to a travelling wave (2.2). The geometrical optics approximation becomes inapplicable since, as a ray family approaches a caustic, the cross-section of the ray tube tends to zero while the amplitude of the ray field grows infinitely. In that case, the wave field is expressed through a more intricate function than a simple exponent (2.2). These special functions come to be known as *functions of wave catastrophes*; they arise naturally in the asymptotic analysis of caustic fields.

In those cases when caustic fields are described by integral representations conforming to, say, a decomposition of a primary field into plane or quasi-plane waves (integral representations of the Fourier type), or into spherical or quasi-spherical waves (representations of the Huygens–Fresnel–Kirchoff type), the functions of wave catastrophes arise as a result of applying the method of stationary phase adapted to the case of several stationary points. The number of stationary points and their mutual location define the type of caustics and corresponding function of wave catastrophes.

Classification of caustics and their respective wave functions is the subject of catastrophe theory put forward basically due to the efforts of Thom and Arnold (1992). One may gain insight into the fundamentals of catastrophe theory by addressing a two-volume book by Arnold, Varchenko and Gussein-Zade (1985, 1988). A brief description of these results as applied to wave problems was given by Kravtsov and Orlov (1983; 1993).

Effective integral field representations generalizing the standard geometrical optics based on the transition to the mixed coordinate-momentum space have been proposed by Maslov (1972) (see also Maslov and Fedoryuk (1981)). The Maslov method is extensively used in applied research concerning electromagnetic wave propagation. The reader may find a concise presentations of this method in works by Kravtsov (1968b) and Kravtsov and Orlov (1993).

One more effective approach is the method of etalon functions or standard integrals (both terms "etalon functions" and "standard integrals" emerge here as synonyms of wave catastrophe functions) which was suggested by Kravtsov (1964a, b) and developed further by Kravtsov and Orlov (1983, 1993). The method of etalon functions abandons the eikonal substitution (2.2) in favor of the field structure characteristic of a given type of caustics. We will use this form of wave field asymptotic representation to analyze how waves mutually transform in a caustic domain. We restrict our consideration to the case of a simple caustic, although it can also be extended to caustics of arbitrary shape.

For a simple caustic it is conventional to express the electric field intensity in terms of the Airy function (integral)

$$\text{Ai}\,(\zeta) = \frac{1}{2\pi} \int_{-\infty}^{\infty} \exp\left[i\left(\zeta t + \frac{t^3}{3}\right)\right] dt. \tag{4.42}$$

However, when constructing uniform asymptotics of a electromagnetic field it is more convenient, instead of (4.42), to use a "dimensional" Airy integral (Kravtsov and Orlov, 1993):

$$\tilde{I}_{Ai}\,(\xi) = \sqrt{\frac{k_0}{2\pi}} \int_{-\infty}^{\infty} \exp[ik_0(\xi s + \frac{s^3}{3})]\, ds, \tag{4.43}$$

related to the conventional Airy integral by a simple relation

$$\tilde{I}_{Ai}\,(\xi) = k_0^{1/6}(2\pi)^{1/2}\,\text{Ai}\,(k_0^{2/3}\xi). \tag{4.44}$$

The uniform asymptotics of an electromagnetic field in the presence of a simple caustic is given by the product of a linear combination of integral $\tilde{I}_{Ai}\,(\xi)$ and its derivative, $\partial \tilde{I}_{Ai}\,(\xi)/\partial\xi$, with the exponential function $\exp(ik_0\chi)$:

$$\mathbf{E} = \left[\mathbf{A}\,\tilde{I}_{Ai}\,(\xi) + (ik_0)^{-1}\mathbf{B}\frac{\partial \tilde{I}_{Ai}\,(\xi)}{\partial\xi}\right] e^{ik_0\chi}. \tag{4.45}$$

Substituting (4.43) into Maxwell's equations (2.3) leads, to low order in k_0, to a system of two equations for the as yet unknown "phase"

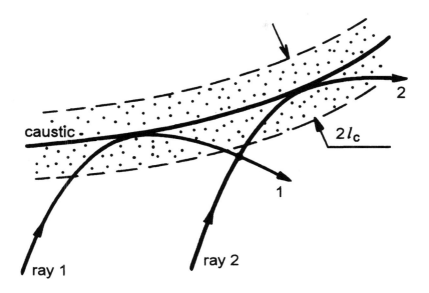

FIGURE 4.3. Trajectories of rays near the caustic. The ray fields $\mathbf{E}_1$ and $\mathbf{E}_2$ allow one to obtain the Airy asymptotics of the wave fields for both isotropic and weakly anisotropic media.

functions ξ and χ. In the zeroth approximation of the quasi-isotropic approach a medium is treated as isotropic, and, hence, ξ and χ obey the same equations as they would in an isotropic medium:

$$\xi(\nabla\xi)^2 + (\nabla\chi)^2 = \varepsilon_0, \qquad (4.46)$$

$$(\nabla\xi\nabla\chi) = 0. \qquad (4.47)$$

According to (4.46) and (4.47) the surfaces $\xi = \text{const}$ and $\chi = \text{const}$ are orthogonal to each other. A natural choice is to bring the surface $\xi = 0$ into coincidence with the caustic surface; there the argument of the Airy integral in (4.43) equals zero. In the domain of caustic shadow, $\xi > 0$, the Airy integral decays exponentially, whereas at $\xi < 0$ it oscillates. The latter corresponds to the interference between two waves, one approaching the caustic and the other moving away from it (Figure 4.3).

Equations in the vector amplitudes $\mathbf{A}$ and $\mathbf{B}$ can be deduced by substituting (4.45) into Maxwell's equations and equating the coefficient at zero with the Airy integral $\tilde{I}_{Ai}(\xi)$ and its derivative $\partial\tilde{I}_{Ai}(\xi)/\partial\xi$. The resultant equations will depart from the analogous equations for isotropic media (Kravtsov and Orlov, 1990, 1993) only in that they contain terms describing a weak anisotropy.

It turns out that the two unknown quantities ξ and χ can be expressed through two geometrical-optical phases φ_1 and φ_2, and two vector amplitudes $\mathbf{A}$ and $\mathbf{B}$, through the geometric-optical amplitudes $\mathbf{E}_1$ and $\mathbf{E}_2$, which form the ray field:

$$\mathbf{E}_{GO} = \mathbf{E}_1\, e^{i\varphi_1} + \mathbf{E}_2\, e^{i\varphi_2}. \tag{4.48}$$

φ_1 and φ_2 are assumed to satisfy the eikonal equation (2.5a), $(\nabla\varphi)^2 = k_0^2$, and the amplitudes $\mathbf{E}_1$ and $\mathbf{E}_2$, to satisfy the QIA equations (2.48). Then

$$k_0\chi = \frac{1}{2}(\varphi_1 + \varphi_2), \quad \frac{2}{3}k_0(-\xi)^{3/2} = \frac{1}{2}(\varphi_1 - \varphi_2) \tag{4.49}$$

and

$$\mathbf{A} = (-\xi)^{1/4}(\mathbf{E}_1 + i\mathbf{E}_2)\frac{e^{i\pi/4}}{\sqrt{2}}, \quad \mathbf{B} = (-\xi)^{-1/4}(\mathbf{E}_1 - i\mathbf{E}_2)\frac{e^{-i\pi/4}}{\sqrt{2}}. \tag{4.50}$$

In the absence of absorption, χ and ξ are real even in the region of caustic shadow $\xi > 0$. Real rays do not penetrate in this region, but two complex rays are present there. Real parts of eikonals φ_1 and φ_2 in the region of shadow coincide both in magnitude and in sign whereas their imaginary parts are of the same magnitude, but have opposite signs. Therefore, behind the caustic

$$\varphi_1 - \varphi_2 = 2i\,\mathrm{Im}\,\varphi_1$$

and

$$\xi = -\left(\frac{3}{2}\frac{\mathrm{Im}\,\varphi_1}{k_0}\right)^{2/3} > 0.$$

The quantity ξ is analytical at the caustic.

While the geometric-optical amplitudes $\mathbf{E}_1$ and $\mathbf{E}_2$ at a caustic have singularities of order $l^{-1/4}$, l being the distance from caustic, the amplitudes $\mathbf{A}$ and $\mathbf{B}$ are finite there. Therefore in the framework of the method of etalon functions the field amplification at the caustic due to ray focusing is described, instead of amplitude growth in ray optics, by special functions which are the Airy functions in the case considered.

With $\chi, \xi, \mathbf{A}$, and $\mathbf{B}$ found the asymptotic solution (4.45) approximately satisfies Maxwell's equations (Kravtsov and Orlov, 1983, 1993). Some distance l_c (see Figure 4.3) away the caustic, which guarantees the transition from special functions $\tilde{I}_{Ai}(\xi)$ and $\partial\tilde{I}_{Ai}(\xi)/\partial\xi$ to their asymptotics, (4.45) reduces to the geometric-optical solution (4.48) with the single distinction that the wave $\mathbf{E}_2 \exp(i\varphi_2)$, which leaves caustic, acquires the caustic phase shift $\Delta\varphi_c = -\pi/2$. This is reflected in the factors $\pm i$ in combinations $\mathbf{E}_1 \pm i\mathbf{E}_2$ entering (4.50).

Thus, the passage of the wave near a caustic affects only the phase of an outgoing wave leaving its polarization unchanged.

Obviously, the polarization state very near the caustic and its peculiarities in the region of the caustic shadow can only be learned from the caustic representation (4.45). To our knowledge, this question is still not analyzed for either isotropic or weakly anisotropic media.

The consideration presented assumes that ray splitting in the vicinity of a caustic can be ignored. In actual fact this requirement extends only to a region very near the caustic with the width

$$\delta l_c = \left(\frac{\rho_{rel}}{2k_0^2}\right)^{1/3}$$

being dependent on the relative curvature ρ_{rel} of the caustic and rays (Kravtsov and Orlov, 1980, 1981, 1990, 1993). Beyond this zone the quasi-isotropic approximation of geometrical optics holds in its standard form.

If the ray splitting cannot be ignored, the combination outlined above of the QIA with the method of etalon functions no longer applies. In this case one needs a more intricate approach to describe wave interaction capable of accounting at the same time of the ray focusing at the caustic and the emergence of waves of other types possibly possessing their own caustics.

Having no such sophisticated theory at our disposal we may only notice that the emergence of the rays of *the second type in a caustic zone formed by the rays of the first type* seems to be a quite rare event. For most applications processes of that kind can be discarded.

5

Optical effects in weakly-anisotropic media

5.1 Tangent conical refraction

5.1.1 GENERAL PICTURE OF THE EFFECT

As is well-known, the conical refraction (internal) takes place on the incidence of a plane wave on a homogeneous crystal if the refractive indices for two types of normal waves with the same direction of $\mathbf{k}$ coincide (Landau and Lifshitz, 1960)

$$n_1(\mathbf{k}, \widehat{\varepsilon}) = n_2(\mathbf{k}, \widehat{\varepsilon}) \, . \tag{5.1}$$

In an anisotropic smoothly inhomogeneous medium the condition (5.1) will not be satisfied everywhere in the volume occupied by a wave, as it is the case in the classical conical refraction effect, but only on the "critical" line (Figure 5.1) where the wave-vector $\mathbf{k}$ is oriented in a proper way with respect to the principal axes of the tensor $\widehat{\varepsilon}(\mathbf{r})$. Evidently, in order to speak of definite magnitude of the wave-vector $\mathbf{k}(\mathbf{r})$ an incident wave should be of definite polarization type, for instance, of the extraordinary one with $\mathbf{k}(\mathbf{r}) = \mathbf{k}_e(\mathbf{r})$.

The critical line appears because the condition (5.1) corresponds to the intersection of two surfaces in the coordinate space:

$$(\mathbf{q_1}, \widehat{\chi}\mathbf{q_1}) - (\mathbf{q_2}, \widehat{\chi}\mathbf{q_2}) = 0 \quad \text{and} \quad (\mathbf{q_1}, \widehat{\chi}\mathbf{q_2}) = 0, \tag{5.1a}$$

where $\widehat{\chi}(\mathbf{r}) = \widehat{\varepsilon}^{-1}(\mathbf{r})$ and vectors $\mathbf{q_1}(\mathbf{r})$ and $\mathbf{q_2}(\mathbf{r})$ constitute a pair of real-valued unit vectors that are orthogonal to each other and to the vector $\mathbf{k}(\mathbf{r})$ which corresponds to the chosen polarization type of the incident wave. As a result, in a smoothly inhomogeneous medium a specific transformation of waves is observed — the tangent conical refraction (Naida, 1979), instead of the ray scattering into the cone (5.1), as is the case with the classical effect of the conical refraction.

As in the example considered above (Section 4) of the quasi-transversal propagation, bringing close together the refractive indices of ordinary and extraordinary waves leads to their mutual transformation

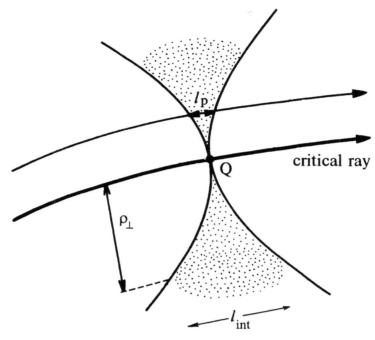

FIGURE 5.1. The region of intensive interaction (dotted) of normal modes in the vicinity of critical point Q where the condition (5.1) of spatial synchronism is satisfied. The characteristic scale l_p of polarization change equals zero at the critical point and increases with increasing the separation between the ray and critical point. The interaction is most effective along the path $l_{int} \sim (k_0 \Delta n)^{-1}$. $\rho_\perp$ is the width of the region of noticeable transformation.

if this is accompanied by a sufficiently rapid polarization transformation of the normal waves. Precisely this situation occurs in the case considered (see Figure 5.2).

A ray intersecting the critical line nor changes its polarization, nor splits. It however changes the polarization type (name) — from the extraordinary to the ordinary one, or conversely (see also Figure 5.3a). On the other hand, a ray (extraordinary or ordinary) passing far enough from the critical line (Figure 5.3b) also does not split, but preserves its polarization type (extraordinary or ordinary) while experiencing the change in the polarization which follows the rotation of the anisotropy axes (see also Figure 5.3b). Finally, for some intermediate distances between a ray and the critical line the wave is able to keep pace with

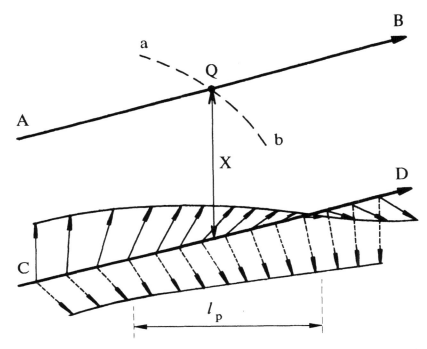

FIGURE 5.2. The change of the polarization of the wave along the ray CD passing close to the critical line (5.1). Solid (dashed) arrows denote the principal axes of the rank 2 tensor $\nu_{\alpha\beta}$ corresponding to the largest (smallest) eigenvalue. The scale l_p of the polarization transformation depends on the distance X to the critical line. On the ray AB intersecting the critical line ab this scale drops to zero.

the rotation of the anisotropy axes only partially. This happens when the length of the region of transformation of the polarization, l_{int}, and the difference of refractive indices, $|n_e - n_o|$ are not so large, namely when $k_0 l_{int} |n_e - n_o| \lesssim 1$. As a result, such rays are subject to splitting, as shown in Figure 5.3b.

Tangent conical refraction may occur in an inhomogeneously deformed crystal, in an inhomogeneously deformed glass whose optical anisotropy is caused by the elastic-optical effect, and also in a moving fluid with an inhomogeneous velocity field, where the anisotropy is due to the Maxwell effect (Landau and Lifshitz, 1960). One may also expect the effect of tangent conical refraction to appear in neodymium glasses subject to heat loads in powerful lasers (Kertes *et al.*, 1970; Foster and

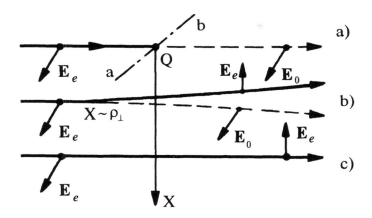

FIGURE 5.3. The polarization transformation of a light wave in the vicinity of the critical line ab as a function of the impact parameter X. Solid lines and solid arrows correspond to extraordinary waves, while dashed ones, to ordinary waves. The rays passing through the critical line ab (a) or far from this line (c) do not practically suffer splitting. The region where splitting is most effective (b) coincides with the region $X \sim \rho_\perp$ where waves interact most effectively. $\mathbf{E}_o$ and $\mathbf{E}_e$ refer to ordinary and extraordinary waves.

Osterlink, 1970). In all cases listed above the dielectric permittivity tensor is real. By dissolving optically-active substances (sugar, etc.) in a moving fluid one may prepare an optically-anisotropic medium with a complex tensor $\hat{\varepsilon}$. In such a medium, the wave transformation will proceed in a rather different manner compared to an optically-neutral medium.

If a birefringent medium with parameters admitting an appreciable wave transformation is placed between crossed polarizers, one may discern the transformation by observing the alterations in the interference pattern on a screen. In Section 5.1.3, we present a relatively simple scheme of an experiment designed to detect the changes that could be introduced by tangent conical refraction.

5.1.2 PARAMETERS OF A LINEAR WAVE TRANSFORMATION

As already noted in Section 3.1, the polarization vectors of electric induction are simultaneously the eigenvectors of the two-dimensional projection $\chi_{\alpha\beta}$ of the tensor $\widehat{\chi} = \widehat{\varepsilon}^{-1}$ in a plane perpendicular to the wave vector $\mathbf{k}$.

In describing the linear transformation of normal waves, we confine ourself to considering the case when the wave arriving at a critical line is either strictly extraordinary or strictly ordinary. We define the ray coordinate s in addition to the above-mentioned (see Section 5.1.1) pair of real unit vectors $\mathbf{q}_1(s)$ and $\mathbf{q}_2(s)$ which are orthogonal to each other and to the wave vector $\mathbf{k}$ on each ray. The directions of the vectors $\mathbf{q}_1(s)$ and $\mathbf{q}_2(s)$ will be chosen so as to provide the invariance of the real part of the tensor component χ_{12} in their basis:

$$\frac{d\left(\operatorname{Re}\chi_{12}\right)}{ds} = 0. \tag{5.1b}$$

In addition to the longitudinal ray coordinate s on the set of rays one may also introduce the transverse coordinates X and Y. If unit vectors $\mathbf{q}_1$ and $\mathbf{q}_2$ are chosen in accordance with the requirement (5.1b), the components χ_{11}, χ_{22} and $\chi_{12} = \chi_{21}^*$ depend on X, Y and s in the following way:

$$\chi_{12} = \chi_{12}(X, Y) = \chi_{21}^*, \quad \chi_{11} = \chi_{11}(X, Y, s),$$
$$\chi_{22} = \chi_{22}(X, Y, s). \tag{5.1c}$$

It seems natural to assume that in the region of intense wave transformation, which is of interest for us, the unit vectors $\mathbf{q}_1(s)$ and $\mathbf{q}_2(s)$ vary sufficiently smoothly at the isotropic inhomogeneity scale l_0.

By virtue of (5.1a), on the critical line the relationships

$$\chi_{11} = \chi_{22}, \qquad \operatorname{Re}\chi_{12} = 0$$

hold. Therefore, $\operatorname{Re}\chi_{12}$ is also equal to zero on the rays passing through the critical line. This implies that on these rays the unit vectors $\mathbf{q}_1$ and $\mathbf{q}_2$ are at the same time the eigenvectors of the two-dimensional part of the tensor $\operatorname{Re}\widehat{\chi}$, and, consequently, specify the directions of normal wave polarizations. The directions of $\mathbf{q}_1$ and $\mathbf{q}_2$ will also be close to those of normal wave polarizations on those parts of rays whose separation from the critical line exceed by far the distance d from the ray to the critical line but is small as compared with l_0. Indeed, let the minimal separation d between the ray and the critical line be small compared to the typical scale l_0 of the medium parameter variability: $d \ll l_0$. Then the distance D between points on the ray and the critical line exceeds d, but when D is small relative to l_0, the quantity $|\chi_{11} -$

$\chi_{22}|$, which grows linearly with the ray coordinate s, may considerably exceed $\mathrm{Re}\,\chi_{12}$, which is small in the vicinity of the critical line. (The imaginary component $\mathrm{Im}\,\chi_{12}$ will be set equal to zero, or, if $\mathrm{Im}\,\chi_{12} \neq 0$, it will be taken into account as a small perturbation in amplitude equations).

An appreciable mutual transformation of ordinary and extraordinary waves occurs along rays satisfying the condition $k_0 l_{\mathrm{int}} \max |n_e - n_o| \lesssim 1$ in the interval of polarization transformation. Apparently, the region of noticeable transformation is confined to the vicinity of the critical line. We designate the width of this region (transverse to rays) by $\rho_\perp$. At its boundary, the scale l_{int} reaches its maximum which will be referred to as the thickness l_{int} of the transformation region.

We confine the analysis to the case of longitudinal localization of the transformation effect, when the longitudinal dimension (along the ray) l_{int} of the intense transformation region is substantially less than the isotropic inhomogeneity scale l_0. Consider the process of wave transformation in the vicinity of a certain point Q on the critical line such that $\chi_{11}(Q) = \chi_{22}(Q), \chi_{12}(Q) = 0$. Let us introduce local Cartesian coordinates s, X, Y in the vicinity of Q where s is a linear coordinate on the beam passing through the critical point Q, and X and Y are the Cartesian coordinates in a plane passing through Q and perpendicular to the ray (we assume here that $s_Q = X_Q = Y_Q = 0$). Then, the components of the tensor $\chi_{\alpha\beta}$ can be treated as linear functions of s:

$$\chi_{11} - \chi_{22} = f(X, Y) + a(X, Y)s, \quad \chi_{12} = g(X, Y). \qquad (5.1d)$$

The picture of the effect shows up even more vividly in the case when the condition of transverse localization ($\rho_\perp \ll l_0$) also holds along with the condition of longitudinal localization ($l_{\mathrm{int}} \ll l_0$). In this case the function $a(X, Y)$ in (5.1d) can be approximated by a constant, while functions $f(X, Y)$ and $g(X, Y)$ can be replaced by linear functions of X and Y:

$$\chi_{11} - \chi_{22} = AX + BY + as, \quad \chi_{12} = CX + DY. \qquad (5.1e)$$

Then one may choose the Y-axis to be in a plane containing the ray and critical line (Figure 5.4). With such a choice, $|X|$ implies the distance d between the ray and the critical line (see Figure 5.2). One may easily see that this choice of X and Y-axes leads to $D = 0$ in Eq. (5.1e). We emphasize that the choice of X and Y-axes in Eq. (5.1e) is to be carried out only *after* the choice of unit vectors $\mathbf{q}_1$ and $\mathbf{q}_2$ in accordance with the condition (5.1b).

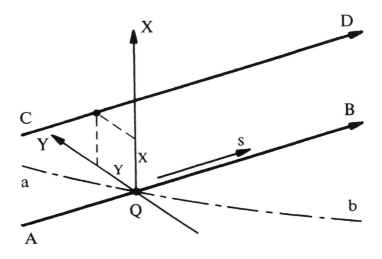

FIGURE 5.4. Local Cartesian coordinate system X, Y, s in the vicinity of the critical point Q. The Y-axis lies in the plane passing through the ray AB and the critical line ab.

The procedure yielding the fields and transformation coefficients was stated in a work by Naida (1979). Practically, it is only slightly different from that described in the preceding chapter as applied to the quasi-transverse wave propagation in a plasma. In both cases the QIA equations are employed. For the longitudinal localization of the conical tangent refraction effect ($l_{int} \ll l_0$) the linearization of the QIA equation coefficients along the ray is accomplished.

In the case of longitudinal localization of the effect, the coefficient of transformation (by intensity) η of an incident linearly polarized wave into a wave of the same type, but with a perpendicular polarization direction, is given by the expression (Naida, 1979)

$$\eta = 1 - \exp(-\pi p/4),\tag{5.2}$$

where

$$p = 2n_o^7 k_0 v^{-1}|\chi_{12}|^2.\tag{5.3}$$

Here $k_0 = \omega/c$, n_0 is the refractive index at the point Q, and v denotes the quantity

$$v = \frac{1}{2}n_0^4 \left| \frac{\partial(\chi_{11} - \chi_{22})}{\partial s} \right|_Q . \tag{5.4}$$

It should be remembered that $\chi_{12} = $ const along each of the rays by virtue of (5.1b). As would be expected, Eq. (5.3) correlates completely with the general formula (2.86).

By virtue of the general formula (2.86) the length of the interval l_p of the polarization transformation is

$$l_p = |\chi_{12}|/|\partial(\chi_{11} - \chi_{22})/\partial s|_Q . \tag{5.4a}$$

Thus, given a real tensor $\hat{\varepsilon}$, for a wave passing through the point Q we obtain $l_p = 0$, since for this ray $\chi_{12} = \chi_{12}(Q) = 0$ (see Figure 5.1). Correspondingly, the transformation coefficient η is equal to zero since $p = 0$. The presence of gyration (Im $\chi_{12} \neq 0$) smooths the effect.

With distance from the critical line the quantities $|\chi_{12}|$ and l_p increase together with the transformation coefficient η which reaches a maximum at $p = 1$ at

$$|\chi_{12}| = \frac{1}{2}\left[n_0^{-3}k_0^{-1}|\partial(\chi_{11} - \chi_{22})/\partial s|_Q\right]^{1/2} .$$

Having substituted this in (5.4a) we may estimate the length of interaction l_{int} (Naida, 1979)

$$l_{int} = l_p|_{p=1} = \frac{1}{2}n_0^{-3/2}k_0^{-1/2}|\partial(\chi_{11} - \chi_{22})/\partial s|_Q^{-1/2} . \tag{5.5}$$

For larger separations from the critical line the quantities l_p and p increase, however the mutual transformation of the ordinary and extraordinary waves dies out. Just for that reason l_{int} given by Eq. (5.5) is the longitudinal (along the ray) dimension of the interaction region depicted in Figure 5.1.

The small size of this dimension compared to l_0 serves as the applicability condition of formula (5.3). If not only the longitudinal l_{int} but also the transverse $\rho_\perp$ dimension of the transformation region is small compared to l_0, then

$$p \approx 2n_0^7 k_0 v^{-1} \left[|\text{Im} \chi_{12}|^2 + X^2 \left| \frac{\partial \text{Re} \chi_{12}}{\partial X} \right|^2 \right]_Q . \tag{5.6}$$

For the parameter $\rho_\perp$ one may conveniently choose the impact parameter $|X|$ for which the light intensity decreases two-fold ($\eta = 1/2$) in the absence of gyration:

$$\rho_\perp = (2\pi^{-1}\ln 2)^{1/2} n_0^{-7/2} k_0^{-1/2} v^{1/2} \left| \frac{\partial \operatorname{Re} \chi_{12}(Q)}{\partial X} \right|^{-1}. \qquad (5.7)$$

This parameter being small with respect to l_0 serves as a condition of transverse localization and, simultaneously, as the applicability condition for the formula (5.6).

5.1.3   THE SIMPLEST SCHEME OF EFFECT OBSERVATION

Figure 5.5 shows the simplest experimental set-up allowing the observation of the effect. It consists of a rectangular glass parallelepiped of length $2L$ and of square base of side a ($L \gg a$), two crossed polarizers P_1 and P_2, the source of monochromatic light, and a screen. The pairs of opposite side faces (AB, CD and BC, AD) can be subjected to distributed loads (of compression) growing linearly from the central cross section to the end-walls, as shown in Figure 5.5. We shall denote the current value of the external pressure by $P(z)$ ($P > 0$ on compression in the direction of orientation of the exit polarizer (x-axis) and $P < 0$ on compression in the perpendicular direction); the maximum value of $|P(z)|$ near the end-walls will be denoted by P_m.

Suppose there is a cylindrical channel of radius R considerably smaller than the width of end-wall face a along the axis of the parallelepiped. Near this channel the glass may be subject to axisymmetric deformations (marginal effects are neglected). These deformations may be created, for example, by applying a constant pressure P_1 from within the channel.

Clearly, if there are no external and internal stresses, the screen will be dark. The same will be observed if there are only external stresses, as shown in Figure 5.5, i. e., at $P_1 = 0$. Lightening due to marginal effects can only be seen in the regions adjacent to the external and internal surfaces.

In the absence of external stresses there are no "critical" points (5.1) on the rays, so there is no linear wave transformation. If external loads ($P_m \neq 0$) are applied, "critical" lines form in the mid-planes $x = 0$ and $y = 0$, and mutual transformation of normal waves occurs close to them. At a given distance r from the center the "critical" points arise when $P_m \geq |\sigma_{rr}^0(r) - \sigma_{\phi\phi}^0(r)|$, where σ_{rr}^0 and $\sigma_{\phi\phi}^0$ are the radial and azimuthal eigenvalues of the stress tensor $\hat{\sigma}^0$ corresponding to the internal (radial) loads. Having originated for a given r at one of the end-walls (at $P_m = |\sigma_{rr}^0(r) - \sigma_{\varphi\varphi}^0(r)|$), the "critical" point, as P_m increases,

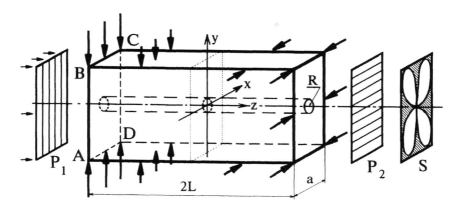

FIGURE 5.5. The simplest scheme for observation of tangent conical refraction. P_1 and P_2 are the crossed polarizers, S is the screen. The points
single out the central cross section of a transparent rectangular parallelepiped
$2L \times a \times a$ with a circular hole of radius R. Solid arrows show the mechanical
loads exerted on the faces of the parallelepiped. The radial loads from within
the channel are not shown. A typical pattern of a cross-like shape is shown
on the screen.

displaces to the central cross section of the parallelepiped. This can
be expressed in another way: the critical line which originates for small
P_m near side-walls of the parallelepiped increases its depth as P_m grows
about. Correspondingly, there is a rise in the modifications brought by
the mutually transforming waves into the interference pattern on the
screen.

It is an easy matter to verify that in the absence of a linear transformation the light intensity at the screen would be

$$I = \frac{1}{2}I_0(1 - \cos 2\psi_1 \cos 2\psi_2 - \sin 2\psi_1 \sin 2\psi_2 \cos \alpha). \qquad (5.8)$$

Here I_0 implies the light intensity at the screen in the absence of an
exit polarizer (nearest to the screen), and ψ_1 and ψ_2 denote the angles
formed by the axis with one of main axes of the tensor $\chi_{\alpha\beta}$ at the
entrance and at exit walls, respectively, of the parallelepiped. Each of
the angles satisfies the equations:

$$\cos 2\psi = \frac{(\sigma_{yy} - \sigma_{xx})}{[(\sigma_{yy} - \sigma_{xx})^2 + 4(\sigma_{xy})^2]^{1/2}},$$

$$\sin 2\psi = \frac{2\sigma_{xy}}{[(\sigma_{yy} - \sigma_{xx})^2 + 4(\sigma_{xy})^2]^{1/2}},$$

(5.9)

where $\hat{\sigma}$ is the total stress tensor accounting for both the external and internal loads. It is implied that the anisotropic constituents of $\hat{\sigma}$ and $\hat{\chi}$ tensors are in proportion:

$$\chi_{yy} - \chi_{xx} = -2n_0^{-3}\kappa(\sigma_{yy} - \sigma_{xx}), \quad \chi_{xy} = -2n_0^{-3}\kappa\sigma_{xy},$$

where n_0 is the averaged refractive index and κ is the elastic-optical coefficient. The quantity α in (5.8) is the phase delay between the extraordinary and ordinary waves gained when passing through the parallelepiped at a distance r from the axis (ray bending is neglected).

If for some r the internal stress is considerably lower that the external one $(|\sigma_{rr}^0(r) - \sigma_{\phi\phi}^0(r)| \ll P_m)$, then, for this r and for ϕ being arbitrary $\sin\psi_1 \ll 1$ and $\sin\psi_2 \ll 1$, i.e., the directions of polarization of a normal wave at the exit and entrance walls are close to those of the x-axis or the y-axis. Recognizing (5.9) we find that at the same time $\cos 2\psi_1 \cos 2\psi_2 \approx -1$, or the deformed glass plays the role of a polarization light guide turning the wave polarization by 90° from the orientation of the entrance polarizer to that of the exit polarizer.

It proves that critical points form on the ray precisely under these conditions $(|\sigma_{rr}^0(r) - \sigma_{\phi\phi}^0(r)| \ll P_m)$ and the polarization light guide just mentioned is capable of that amount of turning only for rays travelling far enough from the critical line and hence from the x and y-axes. There the ratio I/I_0 will be close to unity, in compliance with (5.8). However as one approaches the x and y-axes this ratio will drop to zero and for that reason there will be a dark cross on the screen against a uniformly light background, the cross being the wider, the smaller the ratio $|\sigma_{rr}^0(r) - \sigma_{\phi\phi}^0(r)|/P_m$ (see Figure 5.6). Eqs. (5.2)–(5.7) enable one to calculate the particular parameters of that cross. It should not be confused with the cross appearing according to (5.8) if $P_m = 0$, or, when the external load is absent (Foster and Osterlink, 1970; Kertes et al., 1970):

$$I = I_0 \sin^2(2\phi) \sin^2(\alpha_0/2),$$

(5.10)

ϕ being the azimuthal coordinate of a point on the screen measured as shown in Figure 5.6. α_0 denotes the phase delay between normal waves

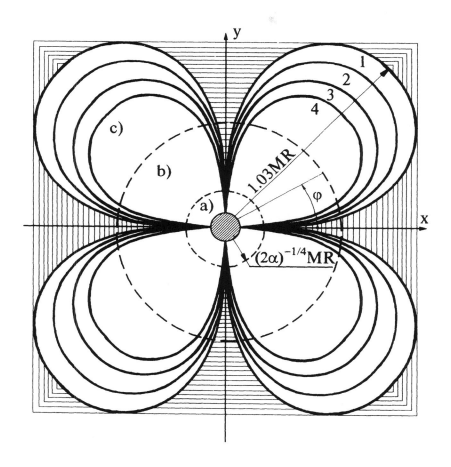

FIGURE 5.6. Isophots corresponding to the tangent conical refraction phenomenon calculated by Eqs. (5.12), (5.13) and (5.15). The intensity is $\frac{1}{2}I_0, \frac{3}{4}I_0, \frac{7}{8}I_0$ and $\frac{15}{16}I_0$ on the curves labelled 1, 2, 3 and 4, respectively. The region where $I < \frac{1}{2}I_0$ is hatched (dark cross). In the region (a) the transformation of waves is negligibly weak; in the region (b) the transverse local effect takes place while in the region (c) it is transversely non-local. The hatched circle in the center corresponds to the channel in the parallelepiped (Figure 5.5)

over the length of glass parallelepiped at $P_m = 0$,

$$\alpha_0(r) = 2k_0 I \kappa [\sigma_{rr}^0(r) - \sigma_{\phi\phi}^0(r)].$$

5.1.4 CALCULATIONS OF LIGHT INTENSITY ON THE SCREEN

To discard the marginal effects we first consider a narrow ring of radius r, well separated from the inner and outer surfaces:

$$R \ll r \ll a.$$

In this case simple analytical expressions follow for the components of the total stress tensor

$$\sigma_{xx} - \sigma_{yy} = (\sigma_{rr}^0 - \sigma_{\phi\phi}^0)\cos 2\phi + P_m z/L,$$

$$\sigma_{xy} = \frac{1}{2}(\sigma_{rr}^0 - \sigma_{\phi\phi}^0)\sin 2\phi. \tag{5.11}$$

Critical points appear where both expressions are equal to zero, i. e., at $z = -(\sigma_{rr}^0(r) - \sigma_{\phi\phi}^0(r))/P_m$, $\phi = 0$, $\phi = \pi$ and $z = (\sigma_{rr}^0(r) - \sigma_{\phi\phi}^0(r))/P_m$, $\phi = \pm\pi/2$, i.e. in the planes $y = 0$ and $x = 0$. For the unit vectors $\mathbf{q}_1$ and $\mathbf{q}_2$ in which (5.2)–(5.7) were written now fit the unit vectors $\mathbf{i}$ and $\mathbf{j}$ of x and y-axes; x and y are then to be used for transverse coordinates X and Y.

Recognizing that the coefficient of transformation of incident linearly polarized wave entering the glass through the polarizer is given by (5.2) the light intensity within the region of transformation is, instead of (5.8),

$$I = I_0(1 - e^{-\pi p/4}), \tag{5.12}$$

while expression (5.3) for the transformation parameter p takes the form

$$p(r, \phi) = 2\kappa k_0 L P_m^{-1}(\sigma_{rr}^0(r) - \sigma_{\phi\phi}^0(r))^2 \sin^2 2\phi. \tag{5.13}$$

The parameter p becomes zero and so does the light intensity I at $\phi = 0, \pi/2, \pi, 3\pi/2$ (on the x and y-axes). The wave passes through these parts of the glass without change of the direction of its polarization though changing its name. On the contrary, as one moves away from the axes and p grows, the ratio I/I_0 tends to unity, which agrees with (5.8).

The estimate (5.5) for the longitudinal dimension h of the region of transformation is now rewritten as

$$l_{\text{int}} \sim L\alpha_P^{-1/2},$$

where $\alpha_P^{-1/2} = \kappa k_0 L P_m$ is the phase shift at the length L created due to an external load P_m in the absence of the internal load, i.e., at $P_1=0$.

Now equating p from (5.13) to the quantity $p_{1/2} = 4\pi^{-1}\ln 2 \approx 0.88$ we obtain the expression for the half-width of the region where $I/I_0 = 1/2$:

$$\rho_\perp = \frac{1}{2}r \left(0.88\alpha_0^{-1} \frac{P_m}{|\sigma_{rr}^0 - \sigma_{\phi\phi}^0|}\right)^{1/2}, \tag{5.14}$$

which conforms to (5.7) as well.

To be specific, we obtain an explicit expression for an isoline of half intensity $r = r_{1/2}(\phi)$ for the case when axisymmetric stresses $\hat{\sigma}(\mathbf{r})$ are created by a uniform pressure P_1 from inside the channel. In this case (Landau and Lifshitz, 1970)

$$\sigma_{rr} - \sigma_{\phi\phi} = -2P_1 R^2/r^2,$$

and (5.13) gives the equation for the sought-for isoline:

$$r = r_{1/2}(\phi) = p_{1/2}^{-1/4}M R|\sin 2\phi|^{1/2}, \quad M = (8k_0 L\kappa P_m^{-1} P_1^2)^{1/4}, \tag{5.15}$$

where $p_{1/2} = 4\pi^{-1}\ln 2$ corresponds to $\eta = 1 - \exp(-\pi p/4) = 1/2$. In particular, when $R < r \ll M R$, i.e., provided r is not too large compared to R, (5.15) rearranges as

$$\rho_\perp(\xi) = \frac{1}{2}M^{-2}\sqrt{p_{1/2}}R^{-2}|\xi|^3, \tag{5.16}$$

where $\rho_\perp$ is the distance from the nearest coordinate axis (either x or y) to the isoline (5.16), with $\xi = |x|$ or $\xi = |y|$. Note that (5.16) may be derived directly from (5.14).

In a similar fashion other isolines $I = wI_0$ can be found. It suffices only to make the replacement in (5.15) and (5.16):

$$p_{1/2} \rightarrow p_w = 4\pi^{-1}\ln\left[1/(1-w)\right]. \tag{5.17}$$

For example, $p_{15/16} = 4p_{1/2}$.

Families of isolines corresponding to (5.15) and (5.17) are shown in Figure 5.6.

For the sake of simplicity Figure 5.6 does not show the perturbations caused by marginal effects at $r \sim a$, and the isolines presented relate to the case when (i) $P_m > |\sigma_{rr}^0 - \sigma_{\phi\phi}^0|$ (the existence of critical points) and (ii) $l_{int} \ll L$ (longitudinal localization of the effect). The latter is equivalent to $\alpha_P \gg 1$ and is realized provided the load is rather high. In this case the former condition also holds, at least, at the point of isophot separated furthest from the center as $|\sigma_{rr}^0 - \sigma_{\phi\phi}^0| \sim \alpha_P^{-1/2} P_m$.

In Figure 5.6 the circumference of radius $M(2\alpha_P)^{-1/4}R$ confines the region a within which $P_m < |\sigma_{rr}^0 - \sigma_{\phi\phi}^0|$, i. e., the transformation is absent. In this region the intensity is given by (5.8) (not shown in Figure 5.6). If $P_m > 2P_1$ this region disappears completely. Eq. (5.8) nearly holds, say, within the lobes of isoline $I = (15/16)I_0$ since (5.8) and (5.12) give nearly coinciding values for I there. In all other parts of Figure 5.6 the picture of tangent conical refraction is not only clearly seen, but also radically contradicts formula (5.8) which assumes the transformation to be absent: this formula requires $I \sim I_0$ everywhere beyond a. Instead, Figure 5.6 shows the region b containing the areas of transversely local effect (the narrow black cross), and c being all the region of non-local transverse effect (the wide black cross passing into a completely black screen).

Now estimate the minimum external loads and the sample length needed to reproduce the isoline (5.15). The total force $F = P_m aL$ exerted on a single parallelepiped face is $\alpha_P a/k_0\kappa$. Taking $\alpha_P = 100$, $a = 3$ cm, $k_0 = 10^7$ m^{-1}, $\kappa = 10^{-12}$ m^2/N we find $F = 3 \times 10^5$ N, i. e. 30 tons. The parallelepiped length $L = \alpha_P/k_0\kappa P_m$. If we take P_m to be a $1/10$ of the strength limit P_s then $L = 10\alpha_P/k_0\kappa P_s$. With $P_s = 3 \times 10^7$ N/m^2 and other parameters as stated previously, $L = 3$ m.

To continue, we now estimate the constraint on the internal pressure P_1. Since P_1 and M are linked by $P_1 \approx (1/3)P_m M^2\alpha_P^{-1/2}$, and since, to smooth marginal effects near the channel, M should be, at least, not less than $M = 3$ we find $P_1 = (1/3)P_m$ with $\alpha_P = 100$. In this case the wave transformation will occur for all $r > R$. With P_1 having a large magnitude, i.e. for large $M^2\alpha_P^{-1/2}$ (5.15) correctly describes only a part of the isoline (see Figure 5.6). Also, increasing M entails increasing the ratio a/R required, to smooth the marginal effects. When the outer faces are close this ratio should not be less than $3M$.

There is also a possibility of simplifying the experiment: one may abandon using limiting cases in which the $\widetilde{\sigma}(\mathbf{r})$ is expressed by elementary functions and work closer to the outer parallelepiped surface (in the region of marginal effects) since there the stresses $\sigma_{\mu\nu}^0$ due to the inner load are appreciably smaller. The dependence $\sigma_{\mu\nu}^0(x,y)$ needed in optical calculations can be established by the interference pattern on the screen in the absence of an external load P_m. Then the half-width b of the region of half intensity could be calculated directly by (5.7). Moreover, one can also abandon the assumption of longitudinal localization, and thus the formula (5.7), and numerically compute the wave transformation and the interference pattern on the screen by directly invoking the differential equations (3.7) for the "tangent conical refraction" field (Naida, 1979). This would allow an order of magnitude reduction in both the sample length and the net external force needed. Thus the "tangent conical refraction" is quite amenable to observations.

5.2   Light propagation in chiral media and inhomogeneous liquid crystals

The photoelastic effect usually leads to a weak optical anisotropy in elastic bodies. Liquid crystals are also characterized by a relatively weak anisotropy. Correspondingly, it seems reasonable to use the quasi-isotropic approximation of geometrical optics to describe light waves in liquid crystals and dielectrics subject to inhomogeneous stresses.

One of the interesting subjects for this type of analysis is light propagation in a chiral medium whose optical axis rotates relative to a ray with a definite spatial period. This subject is interesting in two respects.

First, historically it was the first example which demonstrated the inapplicability of the geometrical optics in the form of independent normal waves in the limit of weak anisotropy (Ginzburg, 1944). Second, a dielectric with a uniformly rotating axis admits an exact solution either of Maxwell's equations (as was derived in the work by Ginzburg (1944) just mentioned above, and also in works by Suvorov, 1972, and Bellustin, 1980) or any approximate equations that could be imagined, including the QIA equations. So in this example, the conditions of applicability of approximate methods can be verified.

Let us imagine that the optical axis OO' of a crystal rotates uniformly with a spatial period Λ relative to the ray propagating in the direction of the z-axis (Figure 5.7), so that the angle φ between the optical axis OO' and the x-axis grows with z as $\varphi(z) = 2\pi z/\Lambda$.

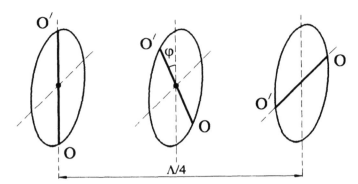

FIGURE 5.7. Uniform rotation of the optical axis OO' in a chiral medium

In the approximation of independent normal waves the polarization vector of the extraordinary electromagnetic wave is tied to the direction of the optical axis and follows its rotation. The vector of the ordinary wave behaves in the same manner. Meanwhile, in the limit of infinitesimal anisotropy $(\mu_1 \sim |\varepsilon_{11} - \varepsilon_{22}| \to 0)$ the electromagnetic field vector should by no means follow the imagined rotation of the virtual (as $\mu_1 \to 0$) optical axis. This contradiction, already noted by Ginzburg (1944) half a century ago, hints at the inapplicability to geometrical optics to the independent normal waves under conditions of weak anisotropy.

Apparently, the inapplicability of geometrical optics in the form of independent normal waves does not immediately imply the inapplicability of all modifications of geometrical optics. For example, in the limit considered, $\mu \to 0$, the quasi-isotropic approximation works reliably. In this approximation it follows, in particular, that in a chiral medium the angle θ in Eq. (2.51) will follow the optical axes, $\theta = \varphi = 2\pi z/\Lambda$, under strong anisotropy when $\delta = \mu_1/\mu = |\Delta\varepsilon|k_0\Lambda \gg 1$, whereas at $\delta \to 0$ the angle θ will not undergo changes: $\theta = \text{const}$.

The applicability of the QIA in a chiral medium hinges on both $\mu_1 \sim |\Delta\varepsilon|$ and $\mu = (k_0\Lambda)^{-1}$ being small:

$$\mu_1 \sim |\varepsilon_{11} - \varepsilon_{22}| \ll 1, \quad \mu \sim (k_0\Lambda)^{-1} \ll 1.$$

The comparison with the exact solution indicates that the error in the QIA is determined by terms of the order of $\mu_1^2 \approx \mu^2$.

The problems of helical wave propagation in inhomogeneous media are of practical importance for the optics of liquid crystals. As shown by Zheleznyakov, Kocharovskii and Kocharovskii (1980), a linear helical wave interaction is feasible in inhomogeneous liquid crystals of cholesteric type characterized by a non-uniform rotation of the optical axes. The work just mentioned also contains the analysis of many other aspects of the question as well as an extensive bibliography.

Another important object which can be solved by the QIA is an inhomogeneous chiral plasma, realized both under laboratory and natural conditions. The interaction of normal waves in a chiral plasma was discussed by Erokhin (1995) and Vacek (1995). The latter work also takes into account nonlinear effects.

5.3   Light polarization in deformed single-mode optical fibers

5.3.1   THE POLARIZATION STATE IN OPTICAL FIBERS

Electromagnetic waves in axisymmetric optical fibers are characterized by a two-fold polarization degeneracy, similar to transverse waves in an isotropic medium. There are many factors lifting the polarization degeneracy in real fibers, such as technological defects in axial symmetry, anisotropy of the fiber material, artificial defects caused by mechanical stresses (photoelasticity), and also bending and torsion of the fibers.

If the propagation constants h_1 and h_2 of two normal waves appearing after the lifting of the degeneracy differ quite strongly ($\Delta h = |h_1 - h_2| \gg 1/l_0$, where l_0 is the characteristic scale of the longitudinal inhomogeneity of the waveguide), the normal waves can be regarded as practically independent. In this case, one may safely suggest that the field polarization in the light guide will be preserved in exciting a wave of a definite type. In the opposite case ($\Delta h \lesssim 1/l_0$), an intense transformation of normal waves occurs. Such a transformation is accompanied by an unstable state of the field polarization in the light guide, which is undesirable in communication systems.

To describe the marked normal wave transformation in light guides it seems natural to resort to equations of the QIA type which would provide a smooth transition both to the polarization degeneracy ($\Delta h \to 0$) and to independent normal waves ($\Delta h \gg 1/l_0$). Here, the role of the anisotropy parameter μ_1 plays the relative difference in the constants of propagation of the two polarization modes, $\mu_1 = \Delta h / l_{\text{int}}$. Taking into account that the parameter $\mu \sim (k_0 l_0)^{-1} \sim (l_{\text{int}} l_0)^{-1}$ preserves its common geometrical sense one can easily verify that the product $\Delta h l_0$ now plays the role of the parameter $\delta = \mu_1/\mu$ and characterizes the wave transformation regime in inhomogeneous parts of light guides:

$$\delta = \frac{\mu_1}{\mu} \sim \frac{\Delta h}{l_{\text{int}}} \frac{1}{(l_{\text{int}} l_0)^{-1}} = l_0 \Delta h.$$

The approach to be stated below was suggested by Kravtsov and Pilipetsky and described in a recent review article by Kravtsov, Naida, and Fuki (1996). It takes into account all basic factors responsible for the polarization state in light guides: bending and torsion of a fiber, weak anisotropy of the material, small deviation from the axial symmetry. Thus, the QIA claims to give a unified consideration of all conceivable polarization effects in light guides, except for, say, an extremely weak scattering on small inhomogeneities.

5.3.2 LOCAL CURVILINEAR COORDINATES

Let $\varepsilon_{ik}(\mathbf{r})$ denote the tensor of fiber dielectric permittivity. The isotropic part of the tensor, $\varepsilon_0 \delta_{ik}$, can, as usual, be obtained by setting $\varepsilon_0 = (1/3) \operatorname{Tr} \varepsilon_{ik}$. The anisotropic part $\nu_{ik} = \varepsilon_{ik} - \varepsilon_0 \delta_{ik}$ is assumed to be small; we shall scale it as $\mu_1 : \mu_1 = \max|\nu_{ik}| \ll 1$.

The fiber axis will be defined as the locus of centroids of the distribution $\delta\varepsilon_0(\mathbf{r}_\perp) = \varepsilon_0(\mathbf{r}_\perp) - 1$ in cross sections. With $\mathbf{r}_\perp$ being the radius-vector in cross section, the center of mass $\mathbf{r}_{\perp c}$ of the distribution $\delta\varepsilon_0(\mathbf{r}_\perp)$ is given by the expression

$$\mathbf{r}_{\perp c} = \frac{\int \delta\varepsilon_0(\mathbf{r}_\perp)\mathbf{r}_\perp \, d^2 r_\perp}{\int \delta\varepsilon_0(\mathbf{r}_\perp) \, d^2 r_\perp}. \tag{5.18}$$

The distribution $\delta\varepsilon_0(\mathbf{r}_\perp) = \varepsilon_0(\mathbf{r}_\perp) - 1$ equals zero outside the light guide, which ensures the convergence of integrals in (5.18). We take the union of points $\mathbf{r}_{\perp c}$ for the waveguide axis $\mathbf{r}_0(\zeta)$, with the distance ζ measured along the axis as a parameter (Figure 5.8). We associate local transverse coordinates ξ, η with the axial line $\mathbf{r}_0(\zeta)$, measuring ξ in the direction of the normal $\mathbf{n}$ and η in the direction of the binormal $\mathbf{b}$ to the axial line. In curvilinear coordinates ξ, η, ζ, the radius vector for an arbitrary point $\mathbf{r}$ can be represented as the sum

$$\mathbf{r}(\xi, \eta, \zeta) = \mathbf{n}\xi + \mathbf{b}\eta + \mathbf{r}_0(\zeta). \tag{5.19}$$

It also seems convenient to introduce polar coordinates (ρ, ϕ) in the

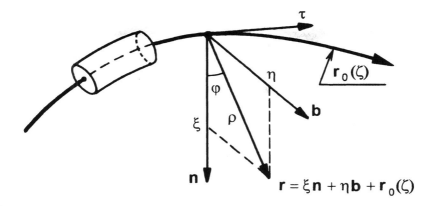

FIGURE 5.8. Local coordinates associated with the axial line $\mathbf{r} = \mathbf{r}_0(\zeta)$ of the optical fiber.

transverse cross-section representing the scalar permittivity $\varepsilon_0(\mathbf{r}_\perp; \zeta)$ as $\varepsilon_0(\rho, \phi; \zeta)$.

The function

$$\varepsilon_s(\rho, \zeta) = \frac{1}{2\pi} \int_0^{2\pi} \varepsilon_0(\rho, \phi; \zeta) \, d\phi, \qquad (5.20)$$

which is dependent on the distance $\rho = (\xi^2 + \eta^2)^{1/2}$ to the axis of the waveguide, will be adopted as the axisymmetric part of the permittivity $\varepsilon_0(\mathbf{r}_\perp; \zeta)$. The small difference

$$\gamma(\rho, \phi; \zeta) = \varepsilon_0(\rho, \phi; \zeta) - \varepsilon_s(\rho, \zeta), \qquad (5.21)$$

will then characterize the departure from axial symmetry. We associate it with the small parameter $\mu_2 = \max(|\gamma|/\varepsilon_0) \ll 1$.

In addition to small parameters μ, μ_1, and μ_2, a new parameter in this problem, $\mu_3 \sim a/l_0 \ll 1$, arises that characterizes the smallness of the radius a of the light guide core as compared to the typical longitudinal scale l_0 of inhomogeneities. One may take the curvature radius of the axial line $R = 1/K_1$ or its torsion radius $T = 1/K_2$ as the scale l_0, so, in fact, $\mu_3' \sim aK_1$ or $\mu_3'' = aK_2$. The smallness of the parameter μ_3 ensures the absence of abrupt bending and twisting of the light guide. At the stage of analytical computations, we shall assume that all parameters μ, μ_1, μ_2, and μ_3 are of the same order of smallness and departing from this we expand the fields in the fiber. However we will be able to analyze the influence of different factors separately afterwards.

5.3.3   ELECTROMAGNETIC FIELD IN AXISYMMETRIC OPTICAL FIBER

An unperturbed system of Maxwell's equations (exact axial symmetry $\varepsilon_0 = \varepsilon_s(\rho)$, absence of anisotropy in the fiber) admits two solutions (e_1, h_1) and (e_2, h_2) corresponding to the main mode LP_{01}. These solutions characterize two polarization states with the same magnitude of the propagation constant h_0 relating to a symmetric distribution of the scalar permittivity $\varepsilon_s(\rho)$

$$e_1 = (e^\perp, 0, e_1^{\parallel}) \exp(ih_0\zeta),$$
$$h_1 = (0, h^\perp, h_1^{\parallel}) \exp(ih_0\zeta),$$
$$\text{(5.22)}$$

$$e_2 = (0, e^\perp, e_2^{\parallel}) \exp(ih_0\zeta),$$
$$h_2 = (-h^\perp, 0, h_2^{\parallel}) \exp(ih_0\zeta).$$
$$\text{(5.23)}$$

Here $e^\perp(\rho)$ and $h^\perp(\rho)$ are the axisymmetric transverse wave functions, which are the same for both polarizations, whereas the longitudinal components have the angular dependence

$$e_1^{\parallel} = e^{\parallel}(\rho) \cos \phi, \quad e_2^{\parallel} = e^{\parallel}(\rho) \sin \phi,$$
$$h_1^{\parallel} = h^{\parallel}(\rho) \sin \phi, \quad h_2^{\parallel} = h^{\parallel}(\rho) \cos \phi.$$
$$\text{(5.24)}$$

The properties of functions $e^\perp(\rho)$, $h^\perp(\rho)$, $e^{\parallel}(\rho)$, and $h^{\parallel}(\rho)$ are described, for example, by Unger (1977), and Snyder and Love (1983). They also present the algorithms of the computation of h_0.

The general solution of Maxwell's equations in a symmetric homogeneous light guide is a superposition of two polarization states (4.22) and (5.23)

$$\mathbf{E} = (\Phi_1 e_1 + \Phi_2 e_2),$$
$$\mathbf{H} = (\Phi_1 h_1 + \Phi_2 h_2).$$
$$\text{(5.25)}$$

5.3.4   MAXWELL'S EQUATIONS IN CURVILINEAR COORDINATES

In a deformed waveguide the vectors of an electromagnetic field can be presented naturally in the local basis $\mathbf{n}$, $\mathbf{b}$, and $\mathbf{t}$, where $\mathbf{t} = \partial \mathbf{r}_0 / \partial \zeta$ is the tangent to the axial line:

$$\mathbf{E} = (E_1 \mathbf{n} + E_2 \mathbf{b} + E_3 \mathbf{t}) \exp\left(i \int h_0(\zeta) d\zeta\right),$$

$$\mathbf{H} = (H_1 \mathbf{n} + H_2 \mathbf{b} + H_3 \mathbf{t}) \exp\left(i \int h_0(\zeta) d\zeta\right),$$

$$(5.26)$$

where the value of h_0 is defined by local characteristics of the fiber.

Let us write the Maxwell equations in curvilinear coordinates ξ, η, and ζ. For that purpose we define the basis $\mathbf{r}_k = \partial \mathbf{r}/\partial x^k$, $\{x^k\} = \{\xi, \eta, \zeta\}$, and a mutual basis $\mathbf{r}^i$, such that $(\mathbf{r}_k \mathbf{r}^i) = \delta_{ik}$. The components of $\mathbf{r}_k$ and $\mathbf{r}^i$ can be expressed through the triple of vectors $\mathbf{n}, \mathbf{b}$ and $\mathbf{t}$:

$$\mathbf{r}_1 = \mathbf{n}, \quad \mathbf{r}_2 = \mathbf{b}, \quad \mathbf{r}_3 = (1 - \xi K_1)\mathbf{t} - \eta K_2 \mathbf{n} + \xi K_2 \mathbf{b},$$

$$\mathbf{r}^1 = \mathbf{n} + \frac{\eta K_2 \mathbf{t}}{1 - \xi K_1}, \quad \mathbf{r}^2 = \mathbf{b} - \frac{\xi K_2 \mathbf{t}}{1 - \xi K_1}, \quad \mathbf{r}^3 = \frac{\mathbf{t}}{1 - \xi K_1},$$

$$(5.27)$$

where K_1 and K_2 are, respectively, the curvature and torsion of the axial line.

Instead of a pair of three-component vectors $\mathbf{E} = \{E_1, E_2, E_3\}$ and $\mathbf{H} = \{H_1, H_2, H_3\}$, it is convenient to introduce a single six-component vector $\mathbf{X} = \{\mathbf{E}, \mathbf{H}\}$. This vector obeys Maxwell's equations which we write in a symbolic form

$$\widehat{A}\mathbf{X} = 0. \tag{5.28}$$

The operator $\widehat{A}$ can be expanded in a series in powers of the small parameter μ:

$$\widehat{A} = \widehat{A}_0 + \widehat{A}_1 + \cdots.$$

The unperturbed part of this operator takes the form

$$
\widehat{A}_0 = \begin{pmatrix}
ik_0\varepsilon_0^0 & 0 & 0 & \vline & 0 & ih_0 & \partial/\partial\eta \\
0 & ik_0\varepsilon_0^0 & 0 & \vline & -ih_0 & 0 & \partial/\partial\xi \\
0 & 0 & ik_0\varepsilon_0^0 & \vline & \partial/\partial\eta & -\partial/\partial\xi & 0 \\
-- & -- & -- & \vline & -- & -- & -- \\
0 & -ih_0 & \partial/\partial\eta & \vline & ik_0 & 0 & 0 \\
ih_0 & 0 & -\partial/\partial\xi & \vline & 0 & ik_0 & 0 \\
-\partial/\partial\eta & \partial/\partial\xi & 0 & \vline & 0 & 0 & ik_0
\end{pmatrix}.
$$
$$(5.29)$$

Small perturbations A_m proportional to μ^m account for the weak anisotropy ν_{ik}, the weak asymmetry γ, and also for torsion and bending of the fiber due to small deviations of the basis $\mathbf{r}_k$ and $\mathbf{r}^i$ (5.27) from a natural trihedron $\mathbf{n}, \mathbf{b}$ and $\mathbf{t}$.

5.3.5   QIA EQUATIONS FOR AMPLITUDES OF POLARIZATION MODES

The system of equations (5.54) can be satisfied by setting

$$\mathbf{X} = \mathbf{X}_0 + \mathbf{X}_1 + \mathbf{X}_2 + \cdots,$$

where $|\mathbf{X}_n| \sim \mu^n |\mathbf{X}_0|$. In the zeroth approximation the field $\mathbf{X}_0$ obeys the equation

$$\widehat{A}_0 \mathbf{X}_0 = 0, \tag{5.30}$$

while all higher approximations are in a series found from the system of equations

$$\widehat{A}_0 \mathbf{X}_1 = -\widehat{A}_1 \mathbf{X}_0,$$

$$\widehat{A}_0 \mathbf{X}_2 = -\widehat{A}_1 \mathbf{X}_1 - \widehat{A}_2 \mathbf{X}_0.$$

A general solution to the zeroth-order approximation (5.30) is expressed through standard fields $\mathbf{X}_{01}$ and $\mathbf{X}_{02}$,

$$\mathbf{X}_0 = \Phi_1 \mathbf{X}_{01} + \Phi_2 \mathbf{X}_{02}. \tag{5.31}$$

These, in turn, are expressed (in local coordinates) through functions $(\mathbf{e_1}, \mathbf{h_1})$ and $(\mathbf{e_2}, \mathbf{h_2})$ according to (5.22)–(5.24). Coefficients Φ_1 and Φ_2 which are undefined as yet follow from the solvability conditions for the equations for the first approximation

$$\widehat{A}_0 \mathbf{X}_1 = -\widehat{A}_1 \mathbf{X}_0 \equiv \mathbf{Z}. \tag{5.32}$$

In the case of two-fold polarization degeneracy, there exist two solvability constraints

$$< \mathbf{ZX}_{01} >= 0 \quad \text{and} \quad < \mathbf{ZX}_{02} >= 0. \tag{5.33}$$

The angular brackets in (5.33) imply a scalar product defined for current purposes as an integral over the wave guide cross section from a conventional scalar product:

$$< \mathbf{A}, \mathbf{B} >= \int (\mathbf{A}, \mathbf{B}) \, d^2 r_\perp.$$

As a result, (5.33) leads to a system of two linked equations for the amplitudes Φ_1 and Φ_2, analogous to the QIA equations. To simplify the analysis we separately consider the contributions coming from geometrical factors (bending and torsion), anisotropy and asymmetry, although in reality they act simultaneously.

5.3.6 THE INFLUENCE OF BENDING AND TORSION

Setting $\nu_{ik} = 0, \gamma = 0$ in the solvability conditions (5.33), i.e. neglecting the anisotropy and asymmetry of the optical fiber we obtain the equations

$$\frac{\partial \Phi_1}{\partial \xi} = -ih_0 K_1^2 a_{\text{eff}}^2 \Phi_1 + K_2 \Phi_2 \,,$$

$$\frac{\partial \Phi_2}{\partial \xi} = -K_2 \Phi_1 + ih_0 K_1^2 a_{\text{eff}}^2 \Phi_2 \,, \tag{5.34}$$

where the parameter

$$a_{\text{eff}} = \left[\frac{\int e^\perp(\rho) h^\perp(\rho) \rho^2 \, d^2 r_\perp}{2 \int e^\perp(\rho) h^\perp(\rho) \, d^2 r_\perp} \right]^{1/2}$$

characterizes an effective radius of the fiber.

Terms with the torsion K_2 in Eqs. (5.34) describe the Rytov rotation of field vectors with respect to the trihedron $\mathbf{n}$, $\mathbf{b}$, and $\mathbf{t}$. A specific feature of this rotation is that it is tied to the axis of the light guide whereas usually the trihedron $\mathbf{n}$, $\mathbf{b}$, and $\mathbf{t}$ is directly associated with the ray in an inhomogeneous medium. The terms with K_1 in (5.34) describe the birefringence in a bend waveguide. For constant curvature and in the absence of torsion (fiber circle), from (5.34) the results follow that were derived by Fung and Lin (1985) on the basis of rigorous electrodynamical considerations of waves in a fiber circle. Corrections to the unperturbed propagation constant h_0 turn out to be different for two polarizations. They are second-order in the parameter $\mu_3' \sim K_1 a_{\text{eff}}$. Although this is somewhat inconsistent, we have artificially attributed them to the first-order terms in Maxwell's equations, and thus obtained the possibility of describing the birefringence due to the fiber bending in the lowest order of perturbation theory.

Results that follow from Eqs. (5.34) compare reasonably with the results of other works (Smith, 1980; Ross, 1984; Itoh, Saitoh and Ohtsuka, 1987) in which the role of bending and torsion was analyzed but using other techniques.

It is noteworthy that the Rytov rotation of the polarization plane was experimentally verified quite recently with the help of light fibers (Tomita and Chiao, 1986; Chiao, Antanamian, Ganga, Jiao, Wilkinson and Nathel, 1988), though studied more than 50 years ago (Rytov, 1938).

5.3.7   THE INFLUENCE OF ELLIPTICITY (ASYMMETRY) OF THE CORE

Expand the asymmetric part γ (5.21) of the permittivity ε_0 in a Fourier series in ϕ

$$\gamma(\rho, \phi, \zeta) = \sum_{m=1}^{\infty} \left[C_m(\rho, \zeta) \cos m\phi + S_m(\rho, \zeta) \sin m\phi \right], \qquad (5.35)$$

and insert it in the solvability condition (5.33). Integrating all terms of this expansion in a combination with symmetric functions $e^{\perp}$, $h^{\perp}$ yields the answer zero while integrating longitudinal components, proportional to $\cos \phi$ and $\sin \phi$ leads to a non-zero result in the combinations with the second azimuthal harmonics C_2 and S_2. As a result, the differential equations for Φ_1 and Φ_2 take the form

$$\frac{\partial \Phi_1}{\partial \zeta} = \frac{ik_0}{2}(\delta_c \Phi_1 + \delta_s \Phi_2),$$

$$\frac{\partial \Phi_2}{\partial \zeta} = \frac{ik_0}{2}(\delta_s \Phi_1 - \delta_c \Phi_2),$$

$$(5.36)$$

where we designated

$$\delta_c = \frac{\int e^{\parallel}(\rho)C_2(\rho,\zeta)\,d^2r_\perp}{4\int e^{\perp}(\rho)h^{\perp}(\rho)\,d^2r_\perp},$$

$$\delta_s = \frac{\int e^{\parallel}(\rho)S_2(\rho,\zeta)\,d^2r_\perp}{4\int e^{\perp}(\rho)h^{\perp}(\rho)\,d^2r_\perp}.$$

(5.37)

From Eqs. (5.36) one may determine the corrections to the propagation constants of two polarization modes, $\Delta h_{1,2} = \pm(\delta_c^2 + \delta_s^2)^{1/2}$. They conform the results of calculations reported by Tjaden (1978), Garth and Pask (1986), and Love and Sammut (1979). An important feature of Eqs. (5.62) is that they enable one to calculate the corrections to the constants of propagation in fibers with arbitrary asymmetry, say for a fiber with two cores. It seems to us that such an approach is simpler and clearer than the existing ones.

5.3.8 THE INFLUENCE OF ANISOTROPY

One distinguishes between two kinds of anisotropy in fibers: "frozen" anisotropy formed in the process of fiber production, and "deformational" anisotropy which occurs due to the action of mechanical stresses, particularly, at bending (Kaminow and Ramaswamy, 1979; Kaminow, 1981; Erickhoff, 1982; Varnham, Payne, Barlow and Birch, 1983).

Incorporating the anisotropy of a light guide material in the perturbation operator $\hat{A}_1$ results in the equations

$$\frac{\partial \Phi_1}{\partial \zeta} = \frac{ik_0}{2}\left[\Phi_1\left(\tilde{\nu}_{nn} + \tilde{\nu}_{nt}^{(1)} + \tilde{\nu}_{tn}^{(1)} + \tilde{\nu}_{tt}^{(11)}\right)\right.$$

$$\left. + \Phi_2\left(\tilde{\nu}_n + \tilde{\nu}_{nt}^{(2)} + \tilde{\nu}_{tb}^{(1)} + \tilde{\nu}_{tt}^{(12)}\right)\right],$$

$$\frac{\partial \Phi_2}{\partial \zeta} = \frac{ik_0}{2}\left[\Phi_1\left(\tilde{\nu}_{bn} + \tilde{\nu}_{tb}^{(1)} + \tilde{\nu}_{tn}^{(2)} + \tilde{\nu}_{tt}^{(12)}\right)\right.$$

(5.38)

$$\left. + \Phi_2\left(\tilde{\nu}_{bb} + \tilde{\nu}_{bt}^{(2)} + \tilde{\nu}_{tb}^{(2)} + \tilde{\nu}_{tt}^{(22)}\right)\right].$$

The coefficients entering the equations are formed from components of the anisotropy tensor $\hat{\nu}$ by the following rules. The coefficients $\tilde{\nu}_{nn}, \tilde{\nu}_{nb}, \tilde{\nu}_{bn},$ and $\tilde{\nu}_{bb}$ are obtained by averaging $\nu_{nn}, \nu_{nb}, \nu_{bn},$ and ν_{bb} with the squared transverse wave function $e^\perp$, for example,

$$\tilde{\nu}_{nn} = \frac{\int \nu_{nn}(\mathbf{r}_\perp)[e^\perp(\rho)]^2\, d^2 r_\perp}{\int e^\perp(\rho)h^\perp(\rho)\, d^2 r_\perp}. \qquad (5.39)$$

Coefficients with a single upper index are the result of averaging with a product of the transverse function $e^\perp(\rho)$ by one of the longitudinal functions $e_1^\|$ or $e_2^\|$, for example

$$\tilde{\nu}_{nt}^1 = \frac{\int \nu_{nt}(\mathbf{r}_\perp)e^\perp(\rho)e_1^\|(\rho,\phi)\, d^2 r_\perp}{\int e^\perp(\rho)h^\perp(\rho)\, d^2 r_\perp}.$$

All these coefficients contain t as one of the lower indices. Finally, components $\tilde{\nu}_{tt}$ with indices 11, 12, or 22 involve ν_{tt} and products of longitudinal functions, as

$$\tilde{\nu}_{tt}^{12} = \frac{\int \nu_{tt}(\mathbf{r}_\perp)e_1^\|(\rho,\phi)e_2^\|(\rho,\phi)\, d^2 r_\perp}{\int e^\perp(\rho)h^\perp(\rho)\, d^2 r_\perp}.$$

For axially symmetric components of the anisotropy tensor, the terms with indices 1, 2, or 12 lead to zero contributions (they contain $\cos\phi$, $\sin\phi$, or $\cos\phi\sin\phi$) and we obtain

$$\frac{\partial \Phi_1}{\partial \zeta} = \frac{ik_0}{2}\left[\Phi_1\left(\tilde{\nu}_{nn} + \frac{1}{2}\tilde{\nu}_{tt}\right) + \Phi_2\tilde{\nu}_{nb}\right],$$

$$\frac{\partial \Phi_2}{\partial \zeta} = \frac{ik_0}{2}\left[\Phi_1\tilde{\nu}_{bn} + \Phi_2\left(\tilde{\nu}_{bb} + \frac{1}{2}\tilde{\nu}_{tt}\right)\right], \qquad (5.40)$$

where $\tilde{\nu}_{nn}, \tilde{\nu}_{nb}, \tilde{\nu}_{bn}$, and $\tilde{\nu}_{bb}$ are, as previously, given by expressions of the type (5.39), and $\tilde{\nu}_{tt}$ stands for

$$\tilde{\nu}_{tt} = \frac{\int \nu_{tt}(\mathbf{r}_\perp)[e^\|(\rho,\phi)]^2\, d^2 r_\perp}{\int e^\perp(\rho)h^\perp(\rho)\, d^2 r_\perp}.$$

From (5.38) and (5.40) it follows that the transformation of field polarization is determined not only by the transverse components of the anisotropy tensor, but also by the longitudinal component, which is also capable of coupling the polarization modes. According to these equations, the propagation constants become changed by $\Delta h \sim k_0|\nu_{ik}|$.

5.3.9 CONDITIONS OF WEAKNESS OF POLARIZATION MODE INTERACTION

Simultaneous action of many factors in differential equations for Φ_1 and Φ_2 is described by a superposition of all the terms mentioned above since the calculations were carried out to first order with respect to μ_i with the only exception for coefficients characterizing the contribution from bending that are proportional to $\mu_3 \sim (K_1 a)^2$.

The advantage of the approach presented is that we consider all feasible factors together, and thus achieve a unified description of various effects that lift the polarization degeneracy.

One of the positive outcomes of this approach is the ability to compare the action of different factors that lift the degeneracy and to find the conditions maintaining a stable polarization in the light guide. In essence we return to the condition

$$\delta \sim \Delta h \, l_0 \gg 1,$$

discussed at the beginning of the section. This condition ensures the weakness of interaction between polarization modes. When the opposite condition, $\delta \lesssim 1$, is implemented, mutual transformation of modes renders the state of polarization in the fiber unstable. The reader can find additional practical details on the problem in works by Kaminow (1981), Payne, Barlow and Ramskov-Hausen (1982), Rashleigh (1983), and Rashleigh and Stolen (1983).

5.4   The optical Magnus effect

To conclude this optical chapter we resort to one more interesting effect which, however, is only indirectly linked with weak optical anisotropy. We are speaking of the optical analog of the Magnus effect.

The essence of the optical Magnus effect is that in an inhomogeneous medium a ray is displaced depending on the degree of circularity of its polarization (Liberman and Zel'dovich, 1992). As a result rays with a right and left circular polarization displace in different directions.

Mentioning the optical Magnus effect, we would like to use it as an additional reason to note a close connection between the anisotropy and inhomogeneity of a medium. In particular, we have already pointed out in Section 4.3.3 that an isotropic medium, possessing a spatial dispersion, has a weak anisotropy if there is an inhomogeneity. In the case of the optical Magnus effect one deals with waves in an isotropic inhomogeneous medium, without spatial dispersion. Although the anisotropy induced by the spatial dispersion does not occur, a weaker effect, the polarization-dependent displacement of rays, is observed.

According to the work of Liberman and Zel'dovich (1992), the optical Magnus effect slightly modifies the ray equation: instead of the first equation in Eqs. (2.12) one should write

$$\frac{d\mathbf{r}}{ds} = \tau - \frac{\sigma}{k_0 n}\left[\tau \ln n_0\right],$$

where τ is the tangent to a ray, and the parameter

$$\sigma = \mathrm{Im}\left(\frac{2E_1^* E_2}{|E_1|^2 + |E_2|^2}\right)$$

characterizes the degree of circularity of the polarization of the field $\mathbf{E}$ (components E_1 and E_2 are the projections of the field on two perpendicular unit vectors). For the right and left circular polarizations $\sigma = +1$ and $\sigma = -1$, respectively, whereas in the intermediate cases $-1 < \sigma < +1$.

The ray displacement mentioned could be interpreted as an interaction of the photon spin (its polarization) with a medium inhomogeneity, i.e. as a sort of a spin–orbit photon interaction in an inhomogeneous medium. In a certain sense the effect is opposite to the Rytov rotation of the polarization plane, in which torsion of a ray influences the field polarization: in the optical Magnus effect the polarization itself influences the ray trajectory.

The optical Magnus effect was observed experimentally in fibers as the speckle-picture displacement (turning) upon replacing the right circular polarization by the left one (Dooghin, Kundikova, Liberman and Zel'dovich, 1992).

5.5   Polarization effects in nonlinear optics

Intense electromagnetic waves in nonlinear isotropic media induce various polarization effects: the self-rotation of the polarization ellipse of a light wave (Maker, Terhune and Savage, 1964), the appearance of nonlinear polarization anisotropy (Akhmanov and Zharikov, 1967; Kaplan, 1983; Law and Kaplan, 1989, 1991a, b), the self-induced rotation of the polarization plane in cubic crystals due to the anisotropy of nonlinear absorption (Dykman and Tarasov, 1977), the appearance of birefringence in a field of intense pumping (Kertes et al., 1970; Foster and Osterlink, 1970).

A wide spectrum of nonlinear polarization phenomena was discussed in a review by Arakelian (1978) devoted to polarization bistability and also in a monograph by Gibbs (1985) devoted to optical bistability.

They include the polarization multistability, polarization instability and chaos, depolarization instability at two-photon absorption, polarization instability in birefringent media, etc. From more recent publications we mention reviews of nonlinear polarization phenomena by Arakelian (1987) and Zheludev (1989) and a paper by Vacek (1995).

In mentioning these works we would like to draw attention to the fact that exactly the QIA ideas are *de facto* applied in analyzing the nonlinear polarization effects. The point is that nonlinear corrections to the tensor of electric permittivity are always considered to be small and are taken into account in equations for field amplitudes as perturbations. If nonlinear corrections are of tensor character, the polarization degeneracy is lifted as the intensity increases. As a result, equations for polarization modes couple to each other and one arrives at a system of nonlinear equations of the QIA. It is this dependence of the anisotropy on amplitudes that leads to the nonlinear polarization effects mentioned above.

6

Geometrical acoustics of weakly anisotropic media

6.1 Quasi-isotropic approximation of geometrical acoustics

6.1.1 BASIC EQUATIONS OF ELASTICITY THEORY

In this section we apply the QIA method to acoustic problems, based on results obtained by Naida (1977b, 1978a). The QIA equations enable matching of the geometrical acoustics of 3D inhomogeneous isotropic media, as elaborated by Levin and Rytov (1956) (their results were later repeated by Karal and Keller, 1959), with the Courant–Lax method of independent normal waves. The latter was successfully applied by Babich (1961) to acoustics of 3D inhomogeneous anisotropic media. Further the results of that work were improved by Cerveny (1972), Cerveny and Psencik (1972), Cerveny, Molotkov and Psencik (1977). The need for such a matching is at present determined by problems arising in seismoacoustics, seismology and ultrasonic non-destructive control of elastic media.

Deformation waves in an inhomogeneous anisotropic medium are described by the equation (Landau and Lifshitz, 1970):

$$\rho \frac{\partial^2 u_\alpha}{\partial t^2} = \frac{\partial \sigma_{\alpha\beta}}{\partial x_\beta}, \tag{6.1}$$

where ρ is the medium density, $\mathbf{u} = (u_1, u_2, u_3)$ is the displacement vector, $\sigma_{\alpha\beta}$ is the stress tensor, connected with the displacements by the formula

$$\sigma_{\alpha\beta} = a_{\alpha\beta\gamma\nu} \frac{\partial u_\gamma}{\partial x_\nu}. \tag{6.2}$$

Here $a_{\alpha\beta\gamma\nu}$ denotes the so-called tensor of elastic moduli ($a_{\alpha\beta\gamma\nu} = a_{\gamma\nu\alpha\beta}$). The summation from 1 to 3 over recurrent Greek indices is implied everywhere in this chapter. In the limiting case of an isotropic medium

$$a_{\alpha\beta\gamma\nu} = a^0_{\alpha\beta\gamma\nu},$$

$$a^0_{\alpha\beta\gamma\nu} = \lambda'\delta_{\alpha\beta}\delta_{\gamma\nu} + \mu'(\delta_{\alpha\gamma}\delta_{\beta\nu} + \delta_{\alpha\nu}\delta_{\beta\gamma}). \tag{6.3}$$

We specially denote the Lamé coefficients by primes to distinguish them from the wavelength λ and the small geometric-optical parameter $\mu = k_0^{-1}l^{-1}$.

For a smoothly inhomogeneous medium, when $\mu = k_0^{-1}l^{-1} \ll 1$ Eqs. (6.1) and (6.2) admit the transition to the substantially simpler equations of geometrical optics.

We shall term a medium a weakly anisotropic one if the relative difference in phase velocities of transverse modes, $\Delta v_\perp / v_\perp$, is small as compared to unity:

$$\mu_1 \sim \frac{\Delta v_\perp}{v_\perp} \ll 1. \tag{6.4}$$

For $\mu_1 = 0$ the geometrical acoustics of isotropic media holds true (Levin and Rytov, 1956; Karal and Keller, 1959), whereas for $\delta = \mu_1/\mu \gg 1$ the geometrical optics in the form of independent normal waves should be invoked (Babich, 1961; Cerveny, 1972; Cerveny and Psencik 1972; Cerveny, Molotkov and Psencik, 1977).

The quasi-isotropic approach describes elastic waves in the intermediate case $0 < \delta \lesssim 1$ and in this way provides a continuous transition from an anisotropic medium to an isotropic one.

6.1.2   QIA EQUATIONS

In a weakly anisotropic medium the Lamé coefficients λ' and μ' can be taken so that the isotropic tensor $a_{\alpha\beta\gamma\nu}$ formed by them differs only slightly from the original tensor $a^0_{\alpha\beta\gamma\nu}$, i. e. that the conditions of weak anisotropy are met:

$$\mu_1 = (\mu' + \lambda')^{-1}\max|\Delta a_{\alpha\beta\gamma\nu}|$$

$$= (\mu' + \lambda')^{-1}\max|a_{\alpha\beta\gamma\nu} - a^0_{\alpha\beta\gamma\nu}| \ll 1. \tag{6.5}$$

We make use of the eikonal substitution

$$\mathbf{u} = \mathbf{U}\exp(-i\omega t + \varphi) \tag{6.6}$$

in Eq. (6.1). Here and henceforth we assume that the medium is stationary and the wave is monochromatic. The generalization to a non-stationary case is straightforward in the absence of dispersion.

Assuming that the amplitude $\mathbf{U}$ and vector $\mathbf{k} = \nabla\varphi$ are of the same degree of smoothness as the parameters λ' and μ', instead of (6.1) and (6.2) we arrive at

$$[(\omega^2\rho - \mu'k^2)U_\alpha - (\lambda' + \mu')(\mathbf{kU})k_\alpha]$$

$$= \Delta a_{\alpha\beta\gamma\nu} k_\beta k_\nu U_\gamma - iX_\alpha + \cdots, \tag{6.7}$$

where

$$\mathbf{X} = (\lambda' + \mu')\big(\nabla(\mathbf{kU}) + \mathbf{k}\,\mathrm{div}\,\mathbf{U}\big) + \mu'(\mathbf{U}\,\mathrm{div}\,\mathbf{k} + 2(\mathbf{k}\nabla)\mathbf{U})$$

$$+(\mathbf{kU})\nabla\lambda' + (\nabla\mu', \mathbf{k})\mathbf{U} + (\nabla\mu', \mathbf{U})\mathbf{k},$$

$$\Delta a_{\alpha\beta\gamma\nu} = a_{\alpha\beta\gamma\nu} - a^0_{\alpha\beta\gamma\nu}.$$

Dots here and in what follows stand for the terms of second or higher orders in μ or μ_1 with respect to the largest of terms accounted for.

We assume that the eikonal φ obeys the equation

$$\omega^2\rho - \mu'(\nabla\varphi)^2 = 0, \tag{6.8}$$

which corresponds to transverse waves in an isotropic medium. The rays can, for example, be calculated from Eqs. (2.12) after replacing ε_0 by ρ/μ'.

Scalarly multiplying Eq. (6.7) by the vectors of the normal $\mathbf{n}$ and binormal $\mathbf{b}$ to the ray, we obtain two equations (with $\mathbf{k} = \nabla\varphi$):

$$(\mathbf{Xn}) + i\Delta a_{\alpha\beta\gamma\nu} n_\alpha k_\beta k_\nu U_\gamma + \cdots = 0,$$

$$(\mathbf{Xb}) + i\Delta a_{\alpha\beta\gamma\nu} b_\alpha k_\beta k_\nu U_\gamma + \cdots = 0. \tag{6.9}$$

Scalarly multiplying (6.7) by the tangent vector $\mathbf{t} = \mathbf{k}/|\mathbf{k}|$ one readily finds an estimate of the transverse component $(\mathbf{Ut})$:

$$|(\mathbf{Ut})| \lesssim \max(\mu_1, \mu)|\mathbf{U}|.$$

This implies that the transverse velocity component is small and to terms of order $\mu \sim (k_0 l)^{-1}$, $\mu_1 \sim \Delta v_\perp/v_\perp$ an approximate solution $\mathbf{U}$ to Eqs. (6.9) might be sought for in the form of a transverse wave, as for an isotropic medium:

$$\mathbf{U} = U_0(\rho\mu')^{-1/4}(Q_n\mathbf{n} + Q_b\mathbf{b}), \tag{6.10}$$

where the normalizing factor U_0 obeys the law (2.37) of energy conservation in a ray tube.

Substitution of (6.10) in (6.9) yields the desired system of approximate equations for Q_n and Q_b:

$$\frac{dQ_n}{ds} + \frac{1}{2} i\omega\rho^{1/2}(\mu')^{-3/2}(\Delta a_{ntnt}Q_n + \Delta a_{ntbt}Q_b) - T^{-1}Q_b = 0\,,$$

$$\frac{dQ_b}{ds} + \frac{1}{2} i\omega\rho^{1/2}(\mu')^{-3/2}(\Delta a_{btnt}Q_n + \Delta a_{btbt}Q_b) + T^{-1}Q_n = 0\,.$$

$$(6.11)$$

where d/ds denotes the derivative along an isotropic ray, T^{-1} is the torsion and indices t, n, b relate to the unit vectors $\mathbf{t}, \mathbf{n}, \mathbf{b}$, respectively, for example, $\Delta a_{ntbt} = \Delta a_{\alpha\beta\gamma\nu} n_\alpha t_\beta b_\gamma t_\nu$. As the anisotropy tends to zero, i. e. at $\Delta a_{\alpha\beta\gamma\nu} \to 0$, Eqs. (6.11) reduce to the equations of Levin and Rytov (1956).

Thus Eqs. (6.7) possess the same structure as the QIA equations for electromagnetic waves. This enables one to extend to acoustics, almost without modification, not only various methods of calculations involving polarization effects but also certain methods of diagnosis of weakly anisotropic media, known in electrodynamics. For instance, by analogy with the method of polarization plasma diagnosis, discussed in Section 4, one may suggest an acoustic method of diagnosis of polarization of anisotropic and pre-stressed media. For the same reason one can anticipate the existence of an acoustic analog of tangent conical refraction (Section 5.1).

If the interaction between normal waves is weak, $\delta = \mu/\mu_1 \gg 1$, Eqs. (6.11) reduce to those for normal waves, obtained by Babich (1961), only up to terms of μ_1 order, as it is generally the case with the QIA. However equations of geometrical acoustics can be also written in the "synthetic" form (3.12a) and (3.12b) (Naida, 1977b, 1978a) which provides an exact transition to the normal waves.

6.2 Geometrical acoustics of an isotropic inhomogeneously stressed medium

6.2.1 ANISOTROPY TENSOR OF STRESSED MEDIUM

Consider acoustic waves propagating in a perfectly elastic and isotropic, but pre-deformed medium. It is implied that this preliminary deformation substantially exceeds the amplitude of the propagating waves and thus that the deformed medium is similar to an anisotropic medium. If one denotes the Lagrangian coordinates associated with medium particles by x_1, x_2, x_3 (they coincide with the Cartesian ones in the absence of deformation), the static displacement vector describing the pre-deformation by $\mathbf{u}^0(\mathbf{r})$, and an additional (small) displacement vector associated with wave motion as $\mathbf{u}(\mathbf{r})$, then the equations linearized with respect to $\mathbf{u}(\mathbf{r})$ attain the form (6.1) except that their coefficients are functions of medium deformation.

As is well-known, in an anisotropic medium

$$a^0_{\alpha\beta\gamma\nu} = \left. \frac{\partial^2 W}{\partial(\frac{\partial w_\alpha}{\partial x_\beta})\,\partial(\frac{\partial w_\gamma}{\partial x_\nu})} \right|_{\partial w_\mu/\partial x_\beta = \partial u^0_\mu/\partial x_\beta} ,$$

where $\mathbf{w} = \mathbf{u}^0 + \mathbf{u}$, and W is the deformation energy per unit of unperturbed volume (Toupin and Bernstein, 1961). Provided the initial disturbances are small we set $a_{\alpha\beta\gamma\nu} = a^0_{\alpha\beta\gamma\nu} + \Delta a_{\alpha\beta\gamma\nu}$, where

$$\Delta a_{\alpha\beta\gamma\nu} = W_{\alpha\beta\gamma\nu\varepsilon\theta} \frac{\partial u^0_\varepsilon}{\partial x_\theta} , \qquad (6.12)$$

$$W_{\alpha\beta\gamma\nu\varepsilon\theta} = \left. \frac{\partial^3 W}{\partial(\frac{\partial w_\alpha}{\partial x_\beta})\,\partial(\frac{\partial w_\gamma}{\partial x_\nu})\,\partial(\frac{\partial w_\varepsilon}{\partial x_\theta})} \right|_{\partial w_\mu/\partial x_\theta = 0}$$

Let us write the energy density W in the form (Landau and Lifshitz, 1970)

$$W = \mu' u^2_{\alpha\beta} + \frac{1}{2}\lambda' u^2_{\alpha\alpha} + \frac{1}{3}A u_{\alpha\beta} u_{\alpha\gamma} u_{\beta\gamma}$$

$$+ B u^2_{\alpha\beta} u_{\gamma\gamma} + \frac{1}{3}C u^3_{\alpha\alpha} + \mathcal{O}(u^4) ,$$

where

$$u_{\alpha\beta} = \frac{1}{2}\left(\frac{\partial w_\alpha}{\partial x_\beta} + \frac{\partial w_\beta}{\partial x_\alpha} + \frac{\partial w_\gamma}{\partial x_\alpha}\frac{\partial w_\gamma}{\partial x_\beta} \right) .$$

The tensor $W_{\alpha\beta\gamma\nu\epsilon\theta}$ then reduces to

$$W_{\alpha\beta\gamma\nu\epsilon\theta} = \mu'[\delta_{\alpha\epsilon}(\delta_{\beta\gamma}\delta_{\nu\theta} + \delta_{\beta\nu}\delta_{\gamma\theta}) + \delta_{\gamma\epsilon}(\delta_{\nu\alpha}\delta_{\beta\theta} + \delta_{\nu\beta}\delta_{\alpha\theta})$$

$$+\delta_{\gamma\alpha}(\delta_{\beta\epsilon}\delta_{\nu\theta} + \delta_{\nu\epsilon}\delta_{\beta\theta})] + \lambda'(\delta_{\alpha\beta}\delta_{\gamma\epsilon}\delta_{\nu\theta} + \delta_{\gamma\nu}\delta_{\alpha\epsilon}\delta_{\beta\theta} + \delta_{\epsilon\theta}\delta_{\alpha\gamma}\delta_{\beta\nu})$$

$$+2A(\delta_{\alpha\gamma}\delta_{\gamma\epsilon}\delta_{\beta\theta})_{(\alpha,\beta)\ (\gamma,\nu)\ (\epsilon,\theta)} + 2C\delta_{\alpha\beta}\delta_{\gamma\nu}\delta_{\epsilon\theta}$$

$$+2B(\delta_{\beta\gamma}\delta_{\alpha\nu}\delta_{\epsilon\theta} + \delta_{\beta\epsilon}\delta_{\alpha\theta}\delta_{\gamma\nu} + \delta_{\alpha\beta}\delta_{\nu\epsilon}\delta_{\gamma\theta})_{(\alpha,\beta)\ (\gamma,\nu)\ (\epsilon,\theta)}\,.$$

$$(6.13)$$

The brackets of the type (α,β) in lower indices denote the symmetrization over the corresponding pair of indices, for instance

$$T_{\alpha\beta\gamma\nu\epsilon\theta(\alpha,\beta)} = \frac{1}{2}(T_{\alpha\beta\gamma\nu\epsilon\theta} + T_{\beta\alpha\gamma\nu\epsilon\theta}).$$

Practically, in a coordinate system where the x_1 axis is parallel to the wave-vector $\mathbf{k}$ one needs to know only $a_{\alpha 1\gamma 1}$ and $W_{\alpha 1\gamma 1\epsilon\theta}$. As it turns out, there are only six different components $W_{\alpha 1\gamma 1\nu\epsilon\theta}$:

$$W_{111111} = W_1 = 6\mu' + 3\lambda' + 2A + 6B + 2C = 6\mu' + 3\lambda' + 4m + 2l\,,$$

$$W_{111122} = W_{111133} = W_2 = \lambda' + 2B + 2C = \lambda' + 2l\,,$$

$$W_{212111} = W_{212122} = W_{112121} = W_{211121} = W_{313111}$$

$$= W_{313133} = W_{113131} = W_{311131}$$

$$= W_3 = 2\mu' + \lambda' + \frac{1}{2}A + B = 2\mu' + \lambda' + m\,,$$

$$W_{212133} = W_{313122} = W_4 = \lambda' + B = \lambda' + m - \frac{1}{2}n\,,$$

$$W_{112112} = W_{211112} = W_{113113} = W_{311113}$$

$$= W_5 = \mu' + \frac{1}{2}A + B = \mu' + m\,,$$

$$W_{213123} = W_{213132} = W_{312132} = W_{312123} = W_6 = \mu' + \frac{1}{4}A = \mu' + \frac{1}{4}n\,,$$

where l, m, and n are the Murnaghan moduli (Murnaghan, 1959):

$$n = A\,, \qquad m = \frac{1}{2}A + B\,, \qquad l = B + C\,.$$

6.2.2   QIA EQUATIONS

Calculating components $\Delta a_{\alpha 1\gamma 1}$ by the formulae (6.12) and substituting them in Eqs. (6.11) we arrive at a particular case of geometric-acoustical equations for a pre-stressed isotropic medium. We shall express the elastic wave field in the form

$$\mathbf{u} = U_0(Q_n\mathbf{n} + Q_b\mathbf{b})(\rho/\mu')^{-1/4}\exp\left\{-i\omega t + i\varphi\right.$$

$$\left. -\frac{i}{2}\omega\int \rho^{1/2}(\mu')^{-3/2}\left[W_3 w_{tt} + \frac{1}{2}(W_3 + W_4)(w_{nn} + w_{bb})\right]ds\right\}.$$

Then, for the amplitudes Q_n and Q_b the system of QIA equations follows

$$\frac{dQ_n}{ds} = -iGQ_n + (T^{-1} - iH)Q_b\,,$$

$$\frac{dQ_b}{ds} = (-T^{-1} - iH)Q_n + iGQ_b\,,$$

(6.14)

where U_0 denotes the parameter satisfying the conservation law along the ray tube, $\operatorname{div}(U_0\mathbf{t}) = 0$, $\mathbf{t} = \mathbf{v}_g/|\mathbf{v}_g| = \mathbf{k}/|\mathbf{k}|$, and

$$G = \frac{\omega\rho^{1/2}}{2(\mu')^{3/2}}\,W_6(w_{nn} - w_{bb})\,,\qquad H = \frac{\omega\rho^{1/2}}{2(\mu')^{3/2}}\,W_6(w_{nb} + w_{bn})\,,$$

$$w_{nn} = n_\alpha n_\beta\frac{\partial u_\alpha^0}{\partial x_\beta}\,,\qquad w_{nb} = n_\alpha b_\beta\frac{\partial u_\alpha^0}{\partial x_\beta}\,,$$

$$w_{bn} = b_\alpha n_\beta\frac{\partial u_\alpha^0}{\partial x_\beta}\,,\qquad w_{bb} = b_\alpha b_\beta\frac{\partial u_\alpha^0}{\partial x_\beta}\,.$$

6.2.3   UNIFORM DEFORMATION OF A HOMOGENEOUS MEDIUM

We illustrate Eqs. (6.14) with simple examples. Assume that a medium is subject to anisotropic compression along x_1, x_2 and x_3-axes. Then the tensor of deformations admits the following form: $w_{ij} = \alpha_i\delta_{ij}$. Since the medium is homogeneous and the ray is straight, the unit vectors $\mathbf{n}$ and $\mathbf{b}$ in (6.14) can be arbitrary provided the orthogonality condition $\mathbf{n} \perp \mathbf{b} \perp \mathbf{t}$ is observed. Let $\mathbf{n}$ and $\mathbf{b}$ align with x_2, x_3 so that $T^{-1} = 0$. Then Eqs. (6.14) give for the velocity v of x_2 linearly polarized transverse waves propagating in the x_1 direction:

$$v = \sqrt{\frac{\mu'}{\rho}} \left\{ 1 + \frac{1}{2\mu'}[(2\mu' + \lambda' + m)(\alpha_1 + \alpha_2) + (\lambda' + m - \frac{1}{2}n)\alpha_3] \right\},$$

where ρ is the density of the undeformed material. Therefore, if the medium is compressed (stretched) in the direction of wave propagation ($\alpha_1 = \alpha \neq 0, \alpha_2 = \alpha_3 = 0$), the velocities of waves of both transverse polarizations coincide:

$$v_I = v_{II} = \sqrt{\frac{\mu'}{\rho}} \left[1 + \frac{1}{2\mu'}(2\mu' + \lambda' + m)\alpha \right]. \tag{6.14a}$$

However, compressing the medium in the transverse direction ($\alpha_1 = 0$) leads to birefringence: the wave polarized in the direction of compression (stretching) ($\alpha_3 = 0, \alpha_2 = \alpha \neq 0$) then acquires the velocity

$$v_I = \sqrt{\frac{\mu'}{\rho}} \left[1 + \frac{1}{2\mu'}(2\mu' + \lambda' + m)\alpha \right],$$

which formally coincides with (6.14a), while the wave polarized perpendicular to the compression ($\alpha_2 = 0, \alpha_3 = \alpha \neq 0$) acquires the velocity

$$v_{II} = \sqrt{\frac{\mu'}{\rho}} \left[1 + \frac{1}{2\mu'}(\lambda' + m - \frac{1}{2}n)\alpha \right].$$

Formula (6.14a) is in agreement with the second of formulae (11) from a work by Hewges and Kelly (1951). One should only bear in mind that we employ Lagrangian, not Eulerian coordinates here.

6.2.4 AXISYMMETRICALLY TWISTED HOMOGENEOUS CYLINDER

As the second example we consider a cylinder of height H and radius R with a fixed axis. Assume that a torque is uniformly applied to its lateral surface, creating shear stresses $\sigma_{n\varphi} = \sigma_0$ in the material near the surface. Let a sound ray pass (not touching the axis) between two points A and B located at the cylinder generators (Figure 6.1). We specify oscillations at the initial point A as

$$\mathbf{u}(A) = (Q_1^A \mathbf{q}_1 + Q_2^A \mathbf{q}_2) \exp(-i\omega t),$$

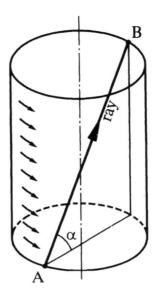

FIGURE 6.1. Acoustic wave propagation in a pre-stressed homogeneous cylinder. The cylinder is subject to tangential forces uniformly distributed over its lateral surface (shown by arrows). The straight acoustic ray is directed at angle α to the cylinder base.

where real unit vectors $\mathbf{q}_1$ and $\mathbf{q}_2$ are perpendicular to each other and to the ray, while $\mathbf{q}_1$ is also perpendicular to the cylinder axis.

Let us find the field $\mathbf{u}$ along the entire ray and, in particular, at point B. For that purpose we introduce x, y, and z axes as shown in Figure 6.1 (so that $x = R_0 = $ const along the ray) and denote the angle between the ray and the x, y-plane by α. We put unit vectors $\mathbf{n} = \mathbf{q}_1$ and $\mathbf{b} = \mathbf{q}_2$ and set $T^{-1} = 0$ in (6.14). Solving the equation of elasticity theory results in the following expressions for the components of the tensor $\widehat{w}$:

$$w_{nn} = -R_0 \frac{y}{r^4} C_0, \quad w_{bb} = -w_{nn} \sin^2 \alpha, \quad w_{tt} = -w_{nn} \cos^2 \alpha,$$

$$w_{nb} + w_{bn} = \frac{y^2 - R_0^2}{r^4} C_0 \sin \alpha,$$

where

$$C_0 = \frac{\sigma_0}{\mu'} R^2, \qquad r^2 = R_0^2 + y^2.$$

By introducing the notations

$$\xi = \frac{y}{R_0}, \quad \delta_1 = \frac{\omega R_0 \rho^{1/2}}{(\mu')^{1/2} \cos \alpha} \frac{W_6}{\mu'} \frac{\sigma_0}{\mu'} \left(\frac{R}{R_0} \right)^2,$$

$$\xi_0 = \frac{y_0}{R_0} = \left(\frac{R^2}{R_0^2} - 1 \right)^{1/2},$$

one may rewrite (6.14) in dimensionless variables:

$$\mathbf{u} = (Q_1 \mathbf{q}_1 + Q_2 \mathbf{q}_2) \exp\left[-i\omega t + i \frac{\omega \rho^{1/2} R_0}{(\mu')^{1/2} \cos \alpha} (\xi + \xi_0) \right.$$

$$\left. + \frac{i}{4} \delta_1 \left(\frac{-1}{1 + \xi_0^2} + \frac{1}{1 + \xi^2} \right) \right],$$

(6.15)

$$\frac{dQ_1}{d\xi} = \frac{-i}{2} \delta_1 \left[-(1 + \sin^2 \alpha) \frac{\xi}{(1 + \xi^2)^2} Q_1 + \sin \alpha \frac{\xi^2 - 1}{(1 + \xi^2)^2} Q_2 \right],$$

$$\frac{dQ_2}{d\xi} = \frac{-i}{2} \delta_1 \left[\sin \alpha \frac{\xi^2 - 1}{(1 + \xi^2)^2} Q_1 + (1 + \sin^2 \alpha) \frac{\xi}{(1 + \xi^2)^2} Q_2 \right].$$

The parameter δ_1 has a clear physical sense: the combination $\delta_1 \arctan \xi_0$ is, by its order of magnitude, a phase delay over the interval AB between normal waves of different polarizations. In the simplest case when $\xi_0 \delta_1 \ll 1$, we obtain

$$\mathbf{u}(B) == (Q_1^B \mathbf{e}_1 + Q_2^B \mathbf{e}_2) \exp\left[-i\omega t + 2i \frac{\omega \rho^{1/2} y_0}{(\mu')^{1/2} \cos \alpha} \right],$$

$$Q_1^B = Q_1^A + i\delta' Q_0^A, \quad Q_2^B = Q_2^A + i\delta' Q_1^A,$$

$$\delta' = \frac{\omega \rho^{1/2} y_0}{(\mu')^{1/2}} \frac{W_6}{\mu'} \frac{\sigma_0}{\mu'}.$$

Thus, if the wave is linearly polarized, then on arriving at B it has a small ($\sim \delta'$) circular component. The exception is the case of $Q_2^A = \pm Q_1^A$ when the wave arriving at B is again linearly polarized (to the first order in $\xi_0 \delta_1$).

Apparently, calculations analogous to those presented above can easily be applied to other acoustic problems, in particular, to non-destructive acoustic control.

6.3   Other possible applications of the QIA to the problems of propagation of elastic waves

In spite of the fact that weakly anisotropic elastic media are widely distributed, attempts to describe their properties based on perturbation theory began to appear quite recently (one of examples was a work by Zhang and Wang (1996)). Meanwhile perturbation theory in terms of the components of the anisotropy tensor not only simplifies calculations, but, more important, makes a description of normal wave interaction effects possible by invoking the QIA approach.

It is noteworthy that such an approach may appear helpful in the acoustics of liquid crystals (Miyano, Ketterson, 1979) and in acoustics of surface waves and piezoelectric crystals (Buryukov, Gulyaev and Plessky, 1995). Also, the QIA may have important implications for the analysis of seismoacoustic signals which, according to recent data (Blackman, Orcutt, Forsyth and Kendall, 1994; Tromp, 1993) pass through inhomogeneous anisotropic regions of the Earth's crust under the oceans.

7

Quasi-isotropic approximation in quantum mechanics

7.1 The Stern–Gerlach effect as birefringence of spinor wave functions in a magnetic field

In this section we compare the QIA with the semiclassical asymptotics of the Pauli equations for spin 1/2 particles in a magnetic field. The case of spin 1/2 and the Pauli equation (as opposed, for instance, to the Dirac equation) are taken here to avoid complex manipulations and demonstrate the essence of novel features brought in by the QIA to this well-studied problem.

There are two approaches to construct the semiclassical approximation for particles with spin: the approach by Pauli (1932) and that by de Broglie (1952). The Pauli approach is based on the assumption that the semiclassical trajectory of a particle is not linked with its magnetic moment. That is quite similar to the early version of the QIA in electrodynamics: in zero order the QIA gives the trajectory (the ray) that does not depend on the spin (the polarization). In his time Pauli did not succeed in deriving a complete (i.e. taking account of polarization) semiclassical asymptotics. This was first done by Galanin (1942). Later the result was rederived, expanded, cleared from computational errors and given a more rigorous treatment by Rubinow and Keller (1963) who obviously knew nothing about the work of Galanin.

Another approach was outlined by de Broglie (1952) who suggested that trajectories of spin 1/2 particles are associated with the magnetic moment already in the zeroth order of the semiclassical approximation. De Broglie intended to supplement the semiclassical formulae by Pauli by a term that would depend on the magnetic moment and external magnetic field. Corresponding additional terms would appear in the eikonal equations and Hamiltonian equations. The latter would imply that particle world lines are dependent on the states of particle spin. This is exactly what occurs in the Stern–Gerlach effect. However de Broglie was unable to find the expression for that additional term.

The program outlined by de Broglie was realized ten years later by Schiller (1962*a,b*), but in a very cumbersome manner. Schiller dealt with five coupled *nonlinear equations* involving three spatial and two spin variables. The complicated character of that system of equations hindered him from isolating a pair of ψ-functions capable of forming a fundamental system of solutions to some *linear* system of equations from all possible solutions. Because of that Schiller's solution did not received a simple and intuitive treatment.

Thus attempts at a rigorous quantum-mechanical description of the Stern–Gerlach have effect always encountered mathematical difficulties. This impelled one either to abandon describing the spatial splitting of the particle beams (Pauli, 1932; Galanin, 1942; Rubinow and Keller, 1963), or to resort to excessively complicated mathematics (de Broglie, 1952; Schiller 1962*a,b*).

Meanwhile the picture of the effect is immediately clarified if one notices that the Stern–Gerlach effect represents the birefringence of a spinor ψ-function by an external magnetic field. Starting from that viewpoint we may quite easily find out solutions for the Pauli and Dirac equations that are similar to those suggested by the electrodynamics of anisotropic inhomogeneous media.

This approach was realized in a paper by Naida and Prudkovskii (1978) which we shall take further. By analogy with the procedure of ray splitting in the electrodynamics of anisotropic media we obtain semiclassical solutions of the Pauli equations that correspond to split rays in the Stern–Gerlach effect, the goal de Broglie was working towards. The solutions admit all the best known solutions as particular cases, in particular, the Stern–Gerlach particle trajectories and solutions of the Pauli–Galanin–Rubinow–Keller type.

7.2   QIA equations for a spinor wave function

Let a beam of non-relativistic polarized spin 1/2 particles (for example, neutrons) enter an inhomogeneous magnetic field (Figure 7.1), as in the experiments by Stern and Gerlach (Stern, 1921; Gerlach and Stern, 1921, 1922*a,b*, 1924). In these experiments, splitting of a beam of polarized particles into two beams with different polarizations was observed. The task is to calculate the polarizations, intensities and phases of these beams. It is noteworthy that neutrons are more convenient as a subject for illustration than Ag or K atoms in $^2S_{1/2}$ states used in the Stern–Gerlach or Frisch–Segré experiments on the reversal

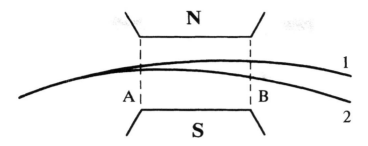

FIGURE 7.1. Splitting of the beam of polarized particles by a magnetic field (the Stern–Gerlach effect). The magnetic field is confined to the interval AB. Trajectories 1 and 2 refer to opposite spin orientations.

of spin states near the neutral point of the magnetic field, the reason being that in atoms we need to account for the magnetic moments of the nucleus, which considerably complicates the picture (Rabi, 1935; Motz and Rose, 1936).

If the neutron wavelength λ is considerably less than the scale l_H of magnetic field inhomogeneity, the solution of this problem reduces to the construction of the appropriate semiclassical solution to the Pauli equation

$$i\hbar \frac{\partial \psi}{\partial t} = \left[-\frac{\hbar^2}{2m} \nabla^2 - \mu_n (\mathbf{H}\boldsymbol{\sigma}) \right] \psi , \qquad (7.1)$$

where m is the neutron mass, μ_n is the neutron magnetic moment $(\mu_n = -0.956|e|\hbar/m_p c)$, $\mathbf{H}$ stands for the magnetic field, and $\boldsymbol{\sigma}$ is a vector made up of Pauli matrices.

In order to obtain a semiclassical solution to the Pauli equation (7.1) we first make the eikonal substitution

$$\psi(x) = A(x) \exp\left[i\varphi(x) \right]. \qquad (7.2)$$

The corresponding eikonal equation with the magnetic moment taken into account is:

$$\det\left(-\hbar\omega + \frac{\hbar^2}{2m}\mathbf{k}^2 - \mu_n\mathbf{H}\boldsymbol{\sigma}\right)$$

$$= \left(-\hbar\omega + \frac{\hbar^2}{2m}\mathbf{k}^2\right)^2 - \mu_n^2\mathbf{H}^2 = 0\,,$$

where $\mathbf{k} = \nabla\varphi$, $\omega = -\partial\varphi/\partial t$. Solving this equation we find

$$\omega = \frac{\hbar}{2m}\mathbf{k}^2 \mp \frac{\mu_n}{\hbar}|\mathbf{H}|\,. \tag{7.3}$$

The corresponding equation for particle trajectories takes the form

$$m\frac{d^2\mathbf{r}}{dt^2} = \pm\mu_n\nabla|\mathbf{H}(\mathbf{r})|\,, \tag{7.4}$$

while the expression for the eikonal φ is:

$$\varphi = \varphi(x_{\text{in}}) + \int_{x_{\text{in}}}^{x}(-\omega\,dt + \mathbf{k}\,d\mathbf{r})\,.$$

Integration is carried out here along the world line (7.4); the vector $\mathbf{k}$ is expressed through $\mathbf{v}$:

$$\mathbf{k} = \frac{m\mathbf{v}}{\hbar}\,. \tag{7.5}$$

The choice of upper (lower) signs in Eqs (7.3) and (7.4) and in subsequent formulae corresponds to spin oriented by (against) the magnetic field. Hence, there are two different eikonals and two different families of the world lines for particles, as de Broglie (1952) suggested.

Clearly, one can afford only to speak of splitting the semiclassical trajectories in the case when the magnetic field $\mathbf{H}$ is strong enough for the Larmor precession to be noticeable compared to the polarization rotation due to the magnetic field inhomogeneity. This implies the inequality

$$\frac{\pi\hbar}{\mu_n|\mathbf{H}|} = T_L \ll \frac{l_H}{v}\,, \tag{7.6}$$

where T_L is the Larmor period, v is for the particle velocity and l_H denotes the field inhomogeneity scale in the direction of the particle trajectory.

In principle, one can always pass to the particle's rest system and solve the Pauli equation there, thus avoiding spatial derivatives. Strictly speaking, the semiclassical approximation employed by Galanin (1942) and later by Rubinow and Keller (1963) represents a rigorous transition to the particle resting frame. However the approach, adopted in works by Majorana (1932), and Frisch and Segré (1933) cannot describe the Stern–Gerlach splitting even if condition (7.6) is met.

Let us derive semiclassical equations for the amplitude $A(x)$. Substituting (7.2) in Eq. (7.1) we find

$$\left(\hbar\omega - \frac{\hbar^2}{2m}\mathbf{k}^2\right)A + i\hbar\left(\frac{\partial A}{\partial t} + \frac{\hbar}{m}\mathbf{k}\nabla A + A\frac{\hbar}{2m}\operatorname{div}\mathbf{k}\right)$$

$$+\mu_{\mathrm{n}}(\mathbf{H}\boldsymbol{\sigma})A = \frac{-\hbar^2}{2m}\nabla^2 A\,. \tag{7.7}$$

Omitting the small right-hand-side in this formula and taking into account (7.5) and (7.3) we deduce the desired semiclassical equation for the amplitude A

$$A = \gamma(x)a(x), \tag{7.8}$$

$$\frac{da}{dt} - i\frac{\mu_{\mathrm{n}}}{\hbar}\big[\mp|\mathbf{H}| + (\mathbf{H}\boldsymbol{\sigma})\big]a = 0\,, \tag{7.9}$$

where $d/dt = \partial/\partial t + \mathbf{v}\nabla$ is the full time derivative along the world line (7.4) and $\gamma(x)$ satisfies the conservation law

$$\frac{\partial\gamma^2}{\partial t} + \operatorname{div}(\mathbf{v}\gamma^2) = 0\,. \tag{7.10}$$

It is easy to see that on altering the phase of the amplitude a in (7.9) in a proper manner, say defining a new amplitude $\tilde{a}$ as

$$\tilde{a} = a\exp\left[\pm i\int_0^t(\mu_{\mathrm{n}}|\mathbf{H}|/\hbar)\,dt\right], \tag{7.11}$$

we convert (7.9) into the Pauli equation in the rest system of a particle moving along the world line (7.4)

$$i\hbar\frac{d\tilde{a}}{dt} = -\mu_{\mathrm{n}}(\mathbf{H}\boldsymbol{\sigma})\tilde{a}. \tag{7.12}$$

Accordingly, along each of the world lines (7.4) the magnetic moment vector

$$\mathbf{M} = \mu_n(a^* \boldsymbol{\sigma} a) = \mu_n(\tilde{a}^* \boldsymbol{\sigma} \tilde{a}) \tag{7.13}$$

satisfies the classical precession equation

$$\frac{d\mathbf{M}}{dt} = 2\mu_n \hbar^{-1} \mathbf{M} \times \mathbf{H}. \tag{7.14}$$

One may easily verify this by substituting (7.13) into (7.14) and taking (7.5) and (7.10) into account.

The eikonal substitution (7.2) is in fact analogous to the quasi-isotropic substitution (as applied to a neutral spin 1/2 particle) that was used by Pauli (1932), Galanin (1942), and also by Rubinow and Keller (1963). The precession equation (7.14) coincides with that obtained by Galanin (1942) and then by Rubinow and Keller (1963) in the non-relativistic classical limit. It was Pauli (1932) who derived the conservation law (7.7).

Combining (7.2), (7.8) and (7.11) gives the desired eikonal substitution

$$\psi(x) = \gamma(x)\,\tilde{a}(x)\,\exp\left\{i\int\left[(-\omega \mp \hbar^{-1}\mu_n|\mathbf{H}|)\,dt + \mathbf{k}\,d\mathbf{r}\right]\right\}, \tag{7.15}$$

which would solve the problem in the de Broglie formulation of the problem. As one can readily see, formula (7.15) refers to the spin direction being parallel or antiparallel to the current vector $\mathbf{H}$, as occurs in the Stern–Gerlach effect.

Thus, the QIA solution presented above involves Eqs. (7.4) for the world lines, the eikonal substitution (7.15) and Eqs. (7.9) for the spin, and differs from the Pauli–Galanin method in the following aspects:

(i) world lines belong to two, rather than one, different types of trajectories which correspond to different particle spin orientations with respect to an external magnetic field;

(ii) phases of ψ-functions contain corrections linked to the magnetic moments.

7.3   Approximation of deformed normal waves

Given arbitrary initial conditions for the amplitude a in (7.9), the second derivative of the amplitude A entering the right-hand side of Eq. (7.7) is of the order

$$|\nabla^2 A| \sim |A| \left(\frac{\mu_n |\mathbf{H}|}{\hbar v} \right)^2 + \cdots, \tag{7.16}$$

and is related to oscillations in a and A with the Larmor frequency. The corresponding error δa associated with solutions of Eq. (7.9) with respect to that of Eq. (7.7) will grow along the ray quicker as $|\mathbf{H}|$ increases:

$$\frac{|\delta a|}{|a|} \sim \frac{\hbar}{m} \int \left(\frac{\mu_n |\mathbf{H}|}{\hbar v} \right)^2 dt . \tag{7.17}$$

The increasing error (7.17) can be eliminated if we choose the initial condition for Eq. (7.9) in the region of strong magnetic field by

$$\left[(\mathbf{H}\sigma) \mp |\mathbf{H}| \right] a = 0 . \tag{7.18}$$

In other words, the initial polarization in the region of the strongest field corresponds to a choice of polarization (with field or against it) which is implicit in Eqs. (7.4) and (7.3). In this case the amplitude $a(x)$ will vary smoothly (without oscillations) in the region of a strong magnetic field and we find instead of the estimate (7.16)

$$|\nabla^2 A| \sim l_H^{-2} |A| + \cdots .$$

Correspondingly, errors accumulated in the region of a strong magnetic field will not then exceed μ_n. On the other hand, in the region of a weak magnetic field the right-hand side of Eq.(7.7) is also small according to estimate (7.16). Therefore, taking the initial condition to Eq. (7.9) to be given by formula (7.18) (in the region of maximal $|\mathbf{H}|$) provides favorable double asymptotes for the solutions to (7.9) both on the side of the strong magnetic field and the weak magnetic field. Naturally, this also promises a high total accuracy satisfying the estimate

$$\frac{|\delta a|}{|a|} \sim \lambda l_B^{-1} , \qquad \lambda = \frac{h}{mv} . \tag{7.19}$$

The critical inhomogeneity scale l_B here implies the same as in the analogous estimates obtained earlier in the context of electrodynamics (see, for example, Sec. 4.1.3). Namely, l_B is the magnitude of the magnetic field inhomogeneity scale l_H at the point of the particle trajectory where

$$l_H \sim \frac{v\hbar}{\mu_n |\mathbf{H}|} \sim v T_L , \tag{7.20}$$

i.e. where l_H coincides to an order of magnitude with the length a particle runs for a Larmor period.

As in the case of electromagnetic waves, we should ascertain the initial condition (7.18) with the help of an iterative procedure to get the highest accuracy (7.19), which as applied to semiclassical equations (7.9) takes the form

$$a = \left(C^1_{(0)} + C^1_{(1)} + C^1_{(2)} + \ldots\right)a_{\pm} + \left(C^2_{(1)} + C^2_{(2)} + \ldots\right)a_{\mp}, \qquad (7.21)$$

where

$$[\mathbf{H}\boldsymbol{\sigma} \mp |\mathbf{H}|]a_{\pm} = 0, \qquad a^*_+ a_+ = a^*_- a_- = 1, \qquad a^*_+ a_- = 0.$$

Coefficient $C^1_{(0)}$ can be found from the equation

$$\frac{dC^1_{(0)}}{dt} + a^*_{\pm}\frac{da_{\pm}}{dt}C^1_{(0)} = 0, \qquad C^1_{(0)}\Big|_{t=t_{in}} = 1, \qquad (7.22)$$

and the other coefficients $C^1_{(n)}$, $C^2_{(n)}$ can be found from the recurrence formulae

$$C^2_{(n)} = \pm\frac{i\hbar}{2\mu_n|\mathbf{H}|}\left[\frac{dC^2_{(n-1)}}{dt} + \left(a^*_{\mp}\frac{da_{\pm}}{dt}\right)C^1_{(n-1)}\right.$$

$$\left. + \left(a^*_{\mp}\frac{da_{\mp}}{dt}\right)C^2_{(n-1)}\right], \qquad (7.23)$$

$$\frac{dC^1_{(n)}}{dt} + a^*_{\pm}\frac{da_{\pm}}{dt}C^1_{(n)} = -a^*_{\pm}\frac{da_{\mp}}{dt}C^2_{(n)}.$$

It is convenient (but not necessary) to set the initial values $C^1_{(n)}(t_{in})$ in the last equation to zero.

One may easily verify that series (7.21) represents an asymptotic expansion of a solution to (7.9) in a region of a strong field. The convergence condition for the series (7.21) is expressed by the inequality

$$l_H \gg v\hbar/\mu_n|\mathbf{H}|, \qquad (7.24)$$

and thus condition (7.20) is a boundary for this inequality.

The ψ-functions constructed in this manner are analogous to deformed normal waves in electrodynamics (see Chapter 3); they correspond exactly to split beams in the Stern–Gerlach effect.

7.4   The trajectory splitting procedure. The spin reversal phenomenon

Provided the error (7.17) is admissible the beam calculations can be accomplished directly in Eq. (7.9) by taking arbitrary upper and lower indices in (7.9), (7.3) and (7.4) and using a given initial condition for the wave function ψ_{in}. The trajectory splitting in this case practically does not occur: it is simply masked by the error estimated by (7.17).

Striving to achieve the upmost accuracy (7.19) we must resort to the procedure of trajectory splitting the essence of which is illustrated by Figure 7.2. We draw two rays through every point of the initial front, corresponding to two types of polarization and reaching the point B where the magnitude of ψ is to be determined. Each of them is constructed by Eq. (7.4) with a respective sign. At the initial points the trajectories have to be perpendicular to the constant phase surface. If each of the trajectories possesses a single maximum of $\mathbf{H}$ (as is assumed in Figure 7.2), then it is necessary to make use of the deformed normal wave approximation in the vicinity of each trajectory and the iterative procedure with a seed (7.21)–(7.23), with the number of steps being given by

$$N \approx \frac{\ln \zeta'}{\ln \left[v\hbar (\mu_n |\mathbf{H}| l_H)^{-1} \right]}, \qquad (7.25)$$

where ζ' is the accuracy required. After that, one should solve Eqs. (7.9) with the result of iterations as an initial condition, each at its own trajectory and with its own initial condition, departing from the initial point to both sides. We shall denote the solutions that result by $b^{\pm}$. One may find the amplitudes b^+ and b^- for each of rays, and then solutions ψ^+ and ψ^- by matching spinors $b^{\pm}$ on the initial front with the initial condition

$$\psi^{\pm}(x) = b^{\pm}(x)\gamma_{\pm}(x) \exp\left\{ i \int_{x_{\text{in}}}^{x} (-\omega_{\pm}\, dt_{\pm} + \mathbf{k}_{\pm}\, d\mathbf{r}_{\pm}) \right\}. \qquad (7.26)$$

Therefore the sought-for approximate solution $\psi(x)$ is

$$\psi(x) = \psi^+(x) + \psi^-(x). \qquad (7.27)$$

Being interested in the interference pattern of the Stern–Gerlach components of the wave, one has to draw a single trajectory from

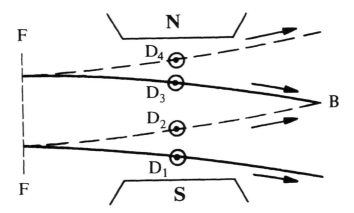

FIGURE 7.2. Construction of wave function at the observation point B by a given initial condition. FF is the initial surface of constant phase. Solid (dashed) lines correspond to trajectories of particles with spin aligned by (against) the magnetic field. $D_{1,2,3,4}$ are the points of local maximum of the magnetic field $|\mathbf{H}^0|$ on the trajectories. Iterative procedures with appropriate seeds are carried out close to these points. The direction of integration is shown by arrows on trajectories, the arrows near them show the directions of motion of the particles. B is the point at which the magnitude of ψ is to be determined

the initial front to the point B considered for each type of polariza-tion(Figure 7.2) and then to implement the procedure described above of constructing the corresponding solutions ψ^+ or ψ^- by Eqs. (7.21)–(7.27) at each of the trajectories. It is this procedure that yields the desired mathematical model of the Stern–Gerlach effect.

If there is more than one local maximum of $|\mathbf{H}|$ on all trajectories or part of them (as is the case in the Frisch–Segré (1933) experiment), the procedure described above of splitting the trajectories and matching the solutions needs to be carried out not only at the initial point, but also at each point of local minimum of the magnetic field $|\mathbf{H}|$ (Figure 7.3), as done in the third and fourth examples in Section 3.3. Accord-ingly, a seed iteration should be constructed in the vicinity of each local maximum of $|\mathbf{H}|$, and the summation in Eq. (7.27) should be accom-plished over multiple (quadruple, octuple, etc.) branching of the rays. As one can see, both Figure 7.3 and the procedure illustrated by it are quite similar to the corresponding procedure in electrodynamics of

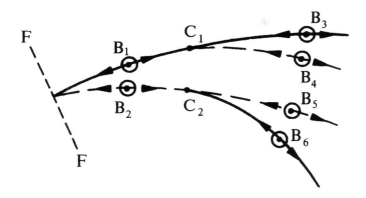

FIGURE 7.3. Construction of a wave function in the presence of of a magnetic field $|\mathbf{H}^0|$ with minima on trajectories (points C_1 and C_2). At these points the trajectories split. At points $B_1, B_2, \ldots, B_6$ the field $|\mathbf{H}^0|$ reaches local maxima. The iterations are constructed there.

birefringent media (Section 3.3).

When dealing with ray splitting it should be remembered that the intensity of the beams splitting off does not exceed $\exp\left[-v\hbar/(\mu_{\mathrm{n}}|\mathbf{H}|l_H)\right]$, and if this factor is small, some splitting can be ignored, for instance when minima of the quantity $|\mathbf{H}|$ are not deep enough.

It is useful to note that for the numerical integration of (7.9) the main difficulty is posed in the interval (7.6) where Eq. (7.9) admits rapidly oscillating solutions. However one may abandon the numerical integration of (7.9) just there by taking advantage of the approximation of deformed normal waves (7.21)–(7.23) at points (Figure 7.3) close to the boundaries of region (7.6). The nearer are these points to the boundary of region (7.6) the easier is the solution of (7.9) in the remaining intervals. However the number of steps of seed iteration required to achieve the desired accuracy is greater. Optimization between these two sources of computational difficulties dominates the choice of matching points in actual fact.

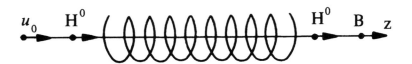

FIGURE 7.4. The passage of a beam of polarized particles along the axis of a solenoid. v_n is the velocity of particles, H^0 is the magnetic field strength. The gauge of the magnetic moment is located at point B.

7.5 Motion of a magnetic moment

Each of the two wave modes $\psi^\pm$ constructed in Section 7.4 possesses a slowly varying and practically non-oscillating amplitude $b^\pm(x)$. Precession of the magnetic moment in this mode comes about due to interference mainly arising from the difference in eikonals corresponding to these modes. Let a neutron beam move along z-axis of an axisymmetric magnetic field from $z = -\infty$ to $z = +\infty$ with an initial velocity v_0 (Figure 7.4) in the framework of a stationary problem. Then the initial wave ψ splits into two waves having, by strength of (7.8), invariant amplitudes along the ray:

$$b^+ = \text{const} \begin{pmatrix} 1 \\ 0 \end{pmatrix} \quad \text{and} \quad b^- = \text{const} \begin{pmatrix} 0 \\ 1 \end{pmatrix}.$$

Hence the interference of waves here stems only from the difference in their eikonals $\phi^\pm$:

$$\phi^+ - \phi^- = \frac{m}{\hbar} \int_{-\infty}^{z} (v^+ - v^-)\, dz$$

$$= \frac{mv_0}{\hbar} \int_{-\infty}^{z} \left[\left(1 + \frac{2\mu_n|\mathbf{H}|}{mv_0^2}\right)^{1/2} - \left(1 - \frac{2\mu_n|\mathbf{H}|}{mv_0^2}\right)^{1/2} \right] dz$$

$$= \int \frac{2\mu_n|\mathbf{H}|}{mv_0^2} \left[1 + \frac{1}{2}\left(\frac{2\mu_n|\mathbf{H}|}{mv_0^2}\right) + \cdots \right] dz. \qquad (7.28)$$

This phase delay can be measured in principle: one may place a detector at an exit from the magnetic field (at point B) and, increasing the magnitude of $|\mathbf{H}|$ from zero to a given value, count the number of oscillations experienced by the magnetic moment of the outgoing neutron beam.

In compliance with the precession equation (7.14), the Galanin–Rubinow–Keller method only gives the first term in squared brackets in (7.28). The appearance of this and the subsequent terms in (7.28) is caused by the fact that split beams move with different velocities as follows from the characteristic equation (7.3) allowing for the magnetic moment. However the first term accounts for this difference not entirely.

Summing up, the asymptotics of the Pauli equations, modified by the QIA method, offers a consistent quantum mechanical picture of the Stern–Gerlach effect as a birefringence of spin ψ-function in an inhomogeneous magnetic field. The role of quantum constraints ensuring the description of the Stern–Gerlach splitting is played by the special initial conditions (7.21)–(7.23) (seed iteration).

Clearly, nothing prevents one from applying the WKB description method developed here for polarized particles to the description of polarized beams of atoms and ions in a magnetic field (needless to say, taking into account the magnetic moment of the nucleus). There are no special difficulties in extending the method to the squared Dirac equations.

7.6 An analogy between the phenomenon of spin reversal and polarization phenomena in propagating electromagnetic waves

Eq. (7.12) describing the polarization of a spin $1/2$ particle in the quasi-isotropic approximation is mathematically equivalent to the QIA equation (2.58) for an electromagnetic wave in a weakly anisotropic medium. Indeed,

$$\mathbf{H}\boldsymbol{\sigma} = \begin{pmatrix} H_z & H_x + iH_y \\ H_x - iH_y & -H_z \end{pmatrix},$$

and Eq. (7.12) can be formally deduced from Eq. (2.66) with due regard for Eqs. (2.62) and (3.7) by resorting to the following change of variables:

$$\begin{pmatrix} \tilde{\Gamma}_1 \\ \tilde{\Gamma}_2 \end{pmatrix} \to \tilde{a}, \qquad k_0 \varepsilon_0^{-1/2} \to \frac{2\mu_a}{\hbar v},$$

$$\gamma_{11} = \frac{\nu_{11} - \nu_{22}}{2} \cong \varepsilon_0^2 \frac{\chi_{22} - \chi_{11}}{2} \to H_z, \qquad (7.29)$$

$$\gamma_{12} - \frac{2i\varepsilon_0^{1/2}}{k_0} T_{\text{eff}}^{-1} = \nu_{12} - \frac{2i\varepsilon_0^{1/2}}{k_0} T_{\text{eff}}^{-1} = -\varepsilon_0^2 \chi_{12} - \frac{2i\varepsilon_0^{1/2}}{k_0} T_{\text{eff}}^{-1} \to H_x + i H_y,$$

where v is the velocity of particles in the beam ($v \ll c$) and μ_a is the magnetic moment of atom with account for the Lande factor.

Departing from this analogy it is natural to expect that in the magneto-optics of spin $1/2$ particles one will encounter the polarization effects similar to the change of the polarization of waves at a quasi-transverse propagation in a magnetized plasma and to the tangent conical refraction of light waves. By way of example we consider an analogy between the classical effect of spin reversal in an alternating magnetic field (Frisch and Segré, 1933) and the phenomenon of tangent conical refraction (section 5.1).

The scheme of the Frisch–Segré experiment is given in Figure 7.5. A thin straight wire W with the current I directed along the y-axis (perpendicular to the plane of the figure) is placed in a uniform external magnetic field $\mathbf{H}_0$ aligned with the z-axis. In the resulting magnetic field $\mathbf{H}(\mathbf{r})$ there exists a neutral point Q. In Figure 7.5, where we show the lines of magnetic field $\mathbf{H}(\mathbf{r})$, the neutral point Q is taken to be the coordinate origin. A narrow beam of potassium atoms is directed perpendicular to both the magnetic field $\mathbf{H}_0$ and the wire W (in the x-axis direction in Figure 7.5). The impact parameter ρ of this ray with respect to the wire W is small compared to the distance x_0 from the wire to the neutral point:

$$\rho \ll x_0 = 2I/H_0. \qquad (7.30)$$

The wave function of an atom in the beam considered obeys the Pauli equation (7.1) but with values of the mass and magnetic moment which are different from those of the neutron. At the initial point A the polarizations of all atoms in the beam are the same, aligned with the field $\mathbf{H}_0$, for example.

The polarizations of the atoms in the beam undergo substantial changes twice. First, on the path AB near the wire the polarization of each atom rotates adiabatically through $180°$ following the direction of the magnetic field and in agreement with Eqs. (7.21)–(7.23). The condition of adiabaticity (7.6) as applied to the path AB takes the form

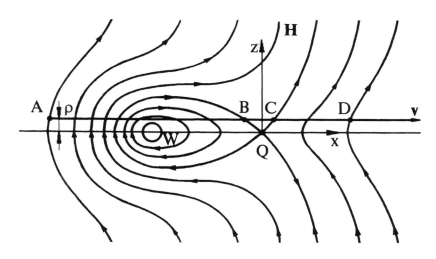

FIGURE 7.5. The pattern of magnetic field lines in the experiment by Frisch and Segré. W is the wire with current, Q is the neutral point. The solid line $ABCD$ depicts the path of polarized spin $1/2$ particles (atoms of potassium).

$$\frac{\hbar v}{\mu_a I} \ll 1,\qquad (7.31)$$

if one sets in Eq. (7.6), by virtue of (7.30),

$$l_H \sim \rho, \qquad H \sim I/\rho. \qquad (7.32)$$

The second substantial change of the polarization of atoms in the beam occurs on the path BC adjacent to the neutral point Q. On this path, the direction of the magnetic field again rotates through $180°$, but in reverse. This rotation takes place across the distance of the order of ρ along the path of the beam. The magnitude of the magnetic field near the neutral point is considerably smaller than close to the wire with current. In actual fact, in the vicinity of the neutral point the components of the total field $\mathbf{H(r)}$ are

$$H_x = \frac{H_0^2}{2I}z, \quad H_z = \frac{H_0^2}{2I}x, \quad H_y = 0. \qquad (7.33)$$

Therefore, the minimum value of the field $|\mathbf{H(r)}|$ along the beam is $H_0^2\rho/2I = 2I\rho/x_0^2$ which is by the factor ρ^2/x_0^2 smaller than in the vicinity of the wire.

Thus, given sufficiently small values of the impact parameter ρ, the process of reorientation of the particle spin may be diabatic. The violation of the condition (7.6) occurs at

$$\frac{\mu_a H_0^2 \rho^2}{\hbar I v} \lesssim 1. \qquad (7.34)$$

Because of this adiabaticity violation an atom with some probability may not follow the change in the orientation of the magnetic field. In this case its polarization at points C and D is the same as it was at point B, i. e. reverse with respect to the initial polarization at the point A. This is the essence of the spin reversal phenomenon.

The opposite process, in which an atom changes the orientation of its spin with the change in the magnetic field direction, occurs with the probability $\eta = 1 - \zeta$. Apparently, the larger the impact parameter, the closer to unity is this probability.

After passing the proximity of the neutral point the trajectories of atoms with opposite spin directions will split as predicted by Eq. (7.4), i.e., in agreement with the Stern–Gerlach effect.

The evolution of the beam polarization on the path BC (in the proximity of the neutral point Q) is most conveniently described by Eq. (7.12) with components of the field $\mathbf{H}$ expressed by Eq. (7.33). This problem was solved by Majorana (1932) who found the following expression for the spin reversal probability:

$$\zeta = \exp\left(-\frac{\pi \mu_a H_0^2 \rho^2}{2\hbar I v}\right). \qquad (7.35)$$

Testing this formula was in fact the aim of the Frisch–Segré experiment. Notice that the argument of the exponent in Eq. (7.35) is of the same order as the left-hand side of the assessment (7.34) derived from qualitative considerations.

Let us follow the analogy between the spin reversal phenomenon and tangent conical refraction. With this aim, we substitute expressions (7.33) for the components of the field $\mathbf{H}$ in Eqs. (7.29) recognizing that the neutral axis Q in spin reversal experiments plays the same role as the critical line (5.1) in the tangent conical refraction, so $z = \rho = \text{const}$ along the beam. It is easy to show that Eq. (7.12) can be formally derived from Eq. (3.7) after the following rearrangement of the variables and constants:

$$\begin{pmatrix} \tilde{\Gamma}_1 \\ \tilde{\Gamma}_2 \end{pmatrix} \to \tilde{a}, \quad k_0 \varepsilon_0^{-1/2} \to \frac{2\mu_a}{hv}, \quad \varepsilon_0^2 \frac{\chi_{22} - \chi_{11}}{2} \to \frac{H_0^2}{2I} s,$$

$$-\varepsilon_0^2 \chi_{12} - \frac{2i\varepsilon_0^{1/2}}{k_0} T_{\text{eff}}^{-1} \to \frac{H_0^2}{2I} \rho. \tag{7.36}$$

As a result of this rearrangement the components of the tensor $\hat{\chi}$ acquire the structure of (5.1d) which leads to the tangent conical refraction. In the framework of the analogy considered, the probability of the spin reversal turns out to be similar to the coefficient of power conversion between normal electromagnetic waves (extraordinary $\rightleftharpoons$ ordinary). Indeed, the coefficient of transformation ζ obeys Eqs. (5.2), (5.3) and (5.4):

$$\zeta = 1 - \eta = \exp(-\pi p/4),$$

$$p = 2n_0^7 k_0 v^{-1} |\chi_{12}|^2, \tag{7.37}$$

$$v = \frac{1}{2} n_0^4 |\partial(\chi_{11} - \chi_{22})/\partial s|, \quad n_0 = \varepsilon_0^{1/2}.$$

After substituting (7.36) in (7.37) one arrives at the expression (7.35) for ζ obtained by Majorana.

It goes without saying that the formal similarity between equations describing the tangent conical refraction in optics and the polarization of spin $1/2$ particles in a magnetic field does not imply that they could be observed under the similar conditions. In reality the observation of the tangent conical refraction assumes the use of wide light beams and polarizers, while in the experiments by Frisch–Segré one deals with narrow beams of atoms whose final polarization states are then revealed with the help of spatial splitting by virtue of the Stern–Gerlach effect.

8

Conclusions

When embarking on the work of this book the authors did not realize fully how multi-faceted the applications of the quasi-isotropic approach are. It applies to vector fields of an arbitrary physical nature: electromagnetic, elastic, spinor. In essence, the QIA concludes the last chapter in the geometrical optics of inhomogeneous media creating the basis for the geometrical optics of weakly anisotropic media. For a long time this branch of geometrical optics was simply absent and there was even a prejudice that geometrical optics has no right to exist in the limit of weak anisotropy. Only with the advent of the QIA has it become clear that weak anisotropy is not an absolute impediment to using the ray method and that the prohibition from applying geometrical optics to weakly anisotropic media only applies to the approaches known previously, i. e., the Rytov method (isotropic media) and the Courant–Lax method (strongly anisotropic media). As a result, a new branch of the ray method has emerged — the geometrical optics of weakly anisotropic media. This branch has already given rise to abundant results and linked together the old branches of geometrical optics that previously existed on their own.

For the most part these results apply to problems of electromagnetic wave propagation in plasmas —, interplanetary, interstellar, solar, ionospheric, laboratory. The subjects for the QIA now also include optical and acoustical phenomena in crystals and deformed isotropic media. The phenomena of tangent conical refraction in optics and elastic wave theory have only been studied theoretically so far. However this work may have important implications for the elaboration of new polarization methods for investigating deformed media.

The quasi-isotropic approximation promises to have many applications in the analysis of the polarization in single-mode fiber light guides. Finally, the quasi-isotropic approximation adequately describes the polarization of particles with spin in inhomogeneous magnetic fields.

Although we tried to collect together all the various types of relevant problems, a number of interesting questions were left untouched.

First and foremost, we have restricted ourselves to the analysis only of wave transformation in a plasma, ignoring the physics of the sources of polarized and partly polarized radio radiation, in particular, in the solar corona. Fortunately, we may address the reader to a book by Zheleznyakov (1977) and to an excellent survey article by Zheleznyakov, Kocharovskii and Kocharovskii (1983) in which the problem is treated in detail.

Further, we were compelled to expound only a small amount of information on the optics of inhomogeneous liquid crystals. The same concerns ultrasonic wave propagation in liquid crystals. We have totally passed in silence over such topics as the transmission of microwaves through weakly anisotropic artificial dielectrics, as well as through forest cover capable of weak birefringent action in the centimeter radiowave band. The questions of wave transformation in weakly deformed dielectric and metallic waveguides of circular and squared cross section subject to polarization degeneracy remained beyond the contents of this book. Ultimately, we have not touched a completely novel physical phenomenon — the birefringence of the vacuum in very strong magnetic fields. It may prove to be an important source of information on matter states under extreme conditions in the Universe.

The authors believe that the concept of the quasi-isotropic approximation, which has an interdisciplinary character, as well as the specific results obtained with the help of QIA will be useful for a wide audience.

References

Akhmanov, S.A. and Zharikov, V.I., 1967. On nonlinear optics of gyrotropic media. *Pis'ma Zhurn.Eksp.Teor.Fiz.*, **6**(5), 644–648. [Engl. transl. in *Pis'ma JETP*, 1967, **6** (5)]

Apresyan, L.A., 1976. Limiting polarization of a wave passing through a thick weakly anisotropic slab with random inhomogeneities.*Astronom. J.*, **53** (1), 53–62.

Apresyan, L.A. and Kravtsov, Yu.A., 1996. *Theory of Radiative Transfer: Wave and Statistical Aspects.* Amsterdam: Gordon and Breach Science Publishers.

Apresyan, L.A. and Kravtsov, Yu.A., 1997. Radiative transfer: new aspects of the old theory. In *Progress in Optics*, (E. Wolf, Ed.), **36**, 181-244. Amsterdam: North Holland.

Apresyan, L.A., Kravtsov, Yu.A., Yashin, Yu.Ya. and Yashnov, V.A., 1976. On the linear transformation of waves in inhomogeneous anisotropic media: 'quasi-degenerate' approximation of geometrical optics. *Izv. VUZ: Radiofizika*, **19**(9), 1296–1304. [Engl. transl. in *Radiophys. and Quantum Electron.*, 1976, **19**]

Akhiezer, A.I. and Berestetskii, V.B., 1965. *Quantum Electrodynamics.* New York: Interscience Publishers.

Arakelian, S.M., 1987. Optical bistability, multistability and instability in liquid crystals. *Uspekhi Fizich. Nauk*, **153**(4), 579–618. [Engl. transl. in *Sov.Phys.-Usp.*, 1987, **30**]

Arnold, V.I., 1992. *Catastrophe Theory*, 3rd edn. Berlin, Heidelberg: Springer Verlag.

Arnold, V.I., Varchenko,A.N. and Gusein-Zade, S.M., 1985. *Singularities of Differentiable Mapping.* Vol.1. Boston: Birkhauser.

Arnold, V.I., Varchenko,A.N. and Gusein-Zade, S.M., 1988. *Singularities of Differentiable Mapping.* Vol.2. Boston: Birkhauser.

Babich, V.M., 1961. Ray method for calculation of the intensity of wave fronts in the case of inhomogeneous anisotropic medium. In *Voprosy dinamicheskoi teorii seismicheskikh voln* [Problems of dynamical theory of seismic waves], No 5, pp.36–46. Leningrad: Leningrad Univ. [in Russian].

Belustin, N.S., 1980. Resonance effects in plasma with helical shear of the magnetic field. *Izv. VUZ: Radiofizika*, **23**(2), 133–142. [Engl. transl. in *Radiophys. and Quantum Electron.*, 1980, **23** (1)].

Berestetskii, V.B., Lifshitz, E.M. and Pitaevskii, L.P., 1971. *Relativistic quantum theory*. Oxford: Pergamon.

Berry, M.V., 1984. Quantal phase factors accompanying adiabatic changes. *Proc. Roy. Soc.* (Lond.), **A392**, 45–67.

Biryukov, S.V., Gulyaev, Yu.V., Krylov, V.V. and Plesskii, V.P., 1995. *Surface Acoustic Waves in Inhomogenous Media*. Berlin, Heidelberg: Springer Verlag.

Blackman, D.K., Orcutt, J.A., Forsyth, D.W. and Kendall, J.- M., 1994. Seismic anisotropic in the mantle beneath an oceanic spreading centre. *Nature*, **366**(6456), 675–678.

Born, M and Wolf, E., 1980. *Principles of Optics*, 6th edn. Oxford: Pergamon.

Born, M., Fock, V. 1928. Beweis des Adiabatensatzes. *ZS. f. Phys.*, **51**, 165–180.

de Broglie, L., 1952. *La Theorie de Particles de Spin 1/2*. Paris.

Budden, K.G., 1952. The theory of the limiting polarization of radio waves reflected from the ionosphere. *Proc.Roy.Soc. A.*, **215**(2), 215–233.

Budden, K.G., 1961.*Radio Waves in the Ionosphere*. Cambridge: Cambridge University Press.

Cerveny, V., 1972. Seysmic rays and ray intensities in inhomogeneous anisotropic media. *Geophys. J.*, **29**(1), 1–13.

Cerveny, V. and Psencik, J., 1972. Rays and travel-times curves in inhomogeneous anisotropic media. *Z. Geophys.*, **38**(3), 565–577.

Cerveny, V., Molotkov, I.A. and Psencik, J., 1977. *Ray Method in Seismology*. Praha: Univerzita Karlova.

Cheng, F.T. and Fung, P.C.W., 1977a. New approach of studying electromagnetic mode coupling in inhomogeneous magnetized plasma. I. Normal incidence. *Astrophys. and Space Sci.*, **49**(2), 367–368.

Cheng, F.T. and Fung, P.C.W., 1977b. The equation of polarization transfer in an inhomogeneous magnetized plasma. I. Formalizm. *Astrophys. and Space Sci.*, **49**(2), 427–442.

Chiao, R.Y. and Wu, Y.-S., 1986. Manifestations of Berry's topological phase for the photon. *Phys. Rev. Lett.*, **57**(8), 933–936.

Chiao, R.Y., Antaramian, A., Ganga, K.M., Jiao, H., Wilkinson, S.R. and Nathel, H., 1988. Observation of a topological phase by means of a nonplanar Mach-Zehnder Interferometer. *Phys. Rev. Lett.*, **60**(13), 1214–1217.

Cohen, M.H., 1960. Magnetoionic mode coupling at high frequencies. *Astrophys. J.*, **131**(3), 664–680.

Courant, R., 1962. Partial Differential Equations. New York: Interscience.

Courant, R. and Lax, P.D., 1956. The propagation of discontinuities in wave motion. *Proc. Nat. Acad. U.S.*, **42**(11), 872–876.

Dooghin, A.V., Kundikova, N.D., Liberman, V.S. and Zel'dovich, B.Ya., 1992. Optical Magnus effect. *Phys. Rev. A.*, **45**(11), 8204–8208.

Dykman, M.I. and Tarasov, G.G., 1977. Absorption saturation and rotation of radiation polarization by local vibrations in cubic crystals. *Zhurn. Eksper. Teor. Fiz.*, **72**(6), 2246–2255. [Engl. transl. in *Sov.Phys. - JETP*, 1977]

Davis, K, 1969. *Ionospheric Radio Waves*. Blaisdell: Wallham.

Eickhoff, W., 1982. Multiple-scattering noise in single-mode fibers. *Opt. Lett.*, **7**(1), 46–48.

Erokhin, N.S. and Moiseev, S.S., 1973. Problems of the theory of linear and nonlinear modes conversion. *Uspekhi Fizich. Nauk*, **109**(2), 225–258. [Engl. transl. in *Sov. Phys. - Uspekhi*, 1973, **16**]

Erokhin, N.S., 1995. On the electromagnetic wave conversion in the weakly inhomogeneous isotropic chiral plasma. PIERS'95, 24–28 July, 1995, Seattle, WA.

Fang, X.-S. and Lin, Z-Q., 1985. Birefringence in curved single-mode optical fibers due to wavequide geometry effect: perturbation analysis. *J. Lightwave Technology*, **3**(3), 789–794.

Foster, J.D. and Osterink, L.H., 1970. Thermal effects in a Nd: YAG laser. *J. Appl. Phys.*, **41**(9), 3656–3663.

Frish, R. and Serge, E., 1933. Uber die Einstellung der Rechtungs Quanelung. *ZS. f. Phys.*, **80**, 610–616.

Fuki, A.A., 1987a. Faradey rotation statistical characteristics of the auroral backscatter. Preprint DEP 4135–B87 (09.06.87). Moscow: VINITI, 16 p.

Fuki, A.A., 1987b. On the solution of the quasi-isotropic approximation equation. Preprint DEP 4136–B87 (09.06.87), Moscow: VINITI, 12 p.

Fuki, A.A., 1987c. The signal depolarization scattered by auroral ionospheric irregularities. Preprint DEP 4593–B87 (24.06.87). Moscow: VINITI, 13 p.

Fuki, A.A., 1988. On the opportunity of electron density measurements in E-layer of polar ionosphere auroral backscatter signals. Preprint no. 881. Moscow: Radiotekhnicheskii Institut AN SSSR, 42 p.

Fuki, A.A., 1990a. Auroral radio-echo polarization characteristics and the lower ionosphere diagnostics. In *Trudy konferentsii 'Radiofizicheskaya Informatika'*. [Proc. Conf. 'Radiophysical Informatics', Moscow, Nov. 27–29, 1990], 128–129. Moscow: Radiotekhnicheskii Institut AN SSSR.

Fuki, A.A., 1990b. Electron density measurement in polar ionosphere with auroral irregularities. Tezisy XVI Vsesoyuznoy Konferentsii po rasprostraneniyu radiovoln [XVI, All-Union Conf. on Radiowave Propagation, Summaries] (Kharkov, Oct. 2–5, 1990). 36–38, Kharkov: Nauchny Sovet AN SSSR po kompleksnoi probleme 'Rasprostranenie radiovoln'.

Fung P.C.W. and Cheng F.T., 1977. New appoach to studying electromagnetic mode coupling in inhomogeneous magnetized plasma. II. Oblique incidence. *Astrophys. and Space Sci.*, **50**(2), 361–381.

Galanin, A.D., 1942. Untersuchung der eigenschaften electronen und mesonenships in der klassischen nahrung. *Journal Physics USSR*, **6**(1,2), 35–47.

Garth, S.J. and Pask, C., 1986. Properties of modes on perturbed fibres. *Electron. Lett.*, **22**(1), 27–28.

Gavrilenko, V.G., Lupanov, G.A. and Stepanov, N.S., 1972. Dynamical effects in plasma. *Izv. VUZ: Radiofizika*, **15**(1), 183–189. [Engl. transl. in *Radiophys. and Quantum Electron.*, 1972, **15**(1)]

Gavrilenko, V.G. and Stepanov, N.S., 1976. Depolarization of electromagnetic waves in turbulent space plasma. *Astron. Zhurn.*, **53**(2),

291–294. [Engl. transl. in *Astron. J.*, 1976, **53**]

Gerlach, W. and Stern, O., 1921. Der experimentelle nachweis des magnetischen moments des silberatoms. *ZS. f. Phys.*, **8**, 110–111.

Gerlach, W. and Stern, O., 1922a. Der experimentelle nachweis der richungsquantelung im magnetfeld. *ZS. f. Phys.*, **9**, 349–352.

Gerlach, W. and Stern, O., 1922b. Das magnetischen moments des silberatoms. *ZS. f. Phys.*, **9**, 353–355.

Gerlach, W. and Stern, O., 1924. Uber die richtungsquantelung im magnetfeld. *Ann. d. Phys.*, **74**(16), 673–699.

Gibbs, H.M., 1985. *Optical Bistability: Controlling Light by Light.* New York: Academic Press.

Ginzburg, V.L., 1970. *The Propagation of Electromagnetic Waves in Plasma.* Oxford: Pergamon.

Ginzburg, V.L., 1944. On the stress study by the optical method. *Zhurn. Tekhnich. Fiz.*, **14**(2), 187–193.

Goryshnik, L.L. and Kravtsov, Yu.A., 1969. Correlation theory of radiowave scattering in polar Ionosphere. *Geomagnetizm i Aeronomia*, **9**(2), 279–285. [Engl. transl. in *Geomagnetism and Aeronomy*, 1969, bf 9(2)]

Goryshnik, L.L., Kravtsov, Yu.A., Tomashuk, L.Ya. and Fomin, B.V., 1969. On the polarization of auroral radioecho. *Geomagnetizm i Aeronomia*, **9**(5), 873–879. [Engl. transl. in *Geomagnetism and Aeronomy*, 1969, **9**(5)]

Gradstein, I.S. and Ryzhik, I.M., 1994. *Table of Integrals, Series and Products.* New York: Academic Press.

Hughes, D.S. and Kelly, J.L., 1951. *Second-order elastic deformation of solids.* New York: Academic Press.

Itoh, K., Saitoh, T. and Ohtsuka, Y., 1987. Optical rotation sensing by the geometric effect of fiber-loop twisting. *J. Light wave Technology*, **5**(7), 916–919.

Kaminow, I.P., 1981. Polarization in Optical Fibers. *IEEE J. Quantum Electron.*, **17**(1), 15–22.

Kaminow, I.P. and Ramaswamy, V., 1979. Single-polarization optical fibers: slab mode. *Appl. Phys. Lett.*, **34**(4), 268–270.

Kantor, I.J., Rai, D.B. and De Mendoca, F., 1970. Behavior of polarization of downcoming radio-waves including transverse magnetic field. *IEEE Trans.*, **AP–19**(2), 246–254.

Kaplan, A.E., 1983. Light-induced nonreciprocity, field invariants and nonlinear eigen-polarizations. *Opt. Lett.*, **8**, 560–562.

Karal, F.C. and Keller J.B., 1959. Elastic wave propagation in homogeneous and inhomogeneous media. *J. Acoust. Soc. Amer.*, **31**(6), 694–705.

Kertes, I., Kononkov, P.G., Senatsky Yu.V. and Chekalin, S.V., 1970. Effect of pump-induced birefringence in neodymium glass on operation of the laser. *Zhurn. Eksper. Teor. Fiziki*, **59**(4), 1115–1124. [Engl. transl. in *Sov. Phys. - JETP*, 1970].

Kharitonova, R.Yu., Fuki, A.A. and Bukatov, M.D., 1989. Circular-polarized signals for auroral hindrance diminishing. In *Radiolokatsyonnye Metody v Radiofizicheskikh Issledovaniyakh* [Radar Methods in Radiophysical Investigations], 76–85. Moscow: RTI AN SSSR [in Russian].

Klyshko, D.H., 1993. Berry geometric phase in oscillatory processes. *Physics - Uspekhi*, **36**(11), 1015–1019.

Kocharovskii, V.V. and Kocharovskii, Vl.V., 1980. On the linear transformation of waves in inhomogeneous media in presence of magnetic lines sheer. *Fizika Plazmy*, **6**(3), 565–576. [Engl. transl. in *Plasma Physics*, 1980, 6]

Korn, G. and Korn, T., 1961. *Mathematical Handbook for Scientists and Engineers*. New York: MGraw-Hill.

Kravtsov, Yu.A., 1964a. Modification of the geometrical optics method, *Izv. VUZ, Radiofizika*, **7**(4), 664–673. [Engl. transl. in *Radiophys. USSR*, **7**(4)]

Kravtsov, Yu.A., 1964b. Asymptotic solution of the Maxwell's equations near caustics. *Izv. VUZ, Radiofizika*, **7**(10), 1049–1056. [Engl. transl. in *Radiophys. USSR*, 1964, **7**(10)]

Kravtsov, Yu.A., 1968a. Quasi-isotropic approximation of geometrical optics. *Doklady AN SSSR*, **183**(1), 74–76 [Engl. transl. in *Sov. Physics - Doklady*, 1968].

Kravtsov, Yu.A., 1968b. Two new asymptotic methods in wave propagation theory through inhomogeneous media. *Sov. Phys. Acoustics*, 1968, **14**(1), 3–19.

Kravtsov, Yu.A., 1988. Rays and caustics as physical objects. In *Progress in Optics*, (E. Wolf, Ed.), v.**26**, 227–348. Amsterdam: North Holland.

Kravtsov, Yu.A., Feizulin, Z.I. and Vinogradov A.G., 1984. *Wave Propagation Through the Earth Atmosphere*. Moscow: Nauka [in Russian].

Kravtsov, Yu.A. and Fuki A.A., 1990. The quasi-isotropic approximation of geometrical optics in problems of ionospheric propagation of radio waves. Tezisy XVI Vsesoyuznoy Konferentsii po rasprostraneniyu radiovoln [XVI, All-Union Conf. on Radiowave Propagation, Summaries] (Kharkov, Oct. 2–5, 1990). 182–185, Kharkov: Nauchny Sovet AN SSSR po kompleksnoi probleme 'Rasprostranenie radiovoln'.

Kravtsov, Yu.A., Kugushev, A.I. and Chernykh, A.V., 1970. Corrections to 'quasi-stationary' value of electric permittivity tensor of smoothly inhomogeneous plasma. *Zhurn. Eksper. Teoret. Fiziki*, **59**(6), 2160–2164. [Engl. transl. in *Sov. Phys. - JETP*, 1970]

Kravtsov, Yu.A. and Naida, O.N., 1976. Linear transformation of electromagnetic waves on a segment of quasi-transversal propagation in 3D inhomogeneous magneto-active plasma. *Sov.Phys. - JETP*, **44**(1), 122–126.

Kravtsov, Yu.A., Naida, O.N. and Fuki, A.A., 1996. Waves in weakly anisotropic 3D inhomogeneous media: quasi-isotropic approximation of geometrical optics. *Physics - Uspekhi*, **39**(2), 129–154.

Kravtsov, Yu.A. and Orlov, Yu.I., 1980. Limits of applicability of the method of geometrical optics and related problems. *Sov. Phys. - Uspekhi*, **23**(11), 750–762. Reprinted in *Geometrical Aspects of Scattering*, Ed. P.L.Marston, SPIE Milestone series, vol.MS89, 1995. Bellingham: SPIE Opt. Eng. Press.

Kravtsov, Yu.A. and Orlov, Yu.I., 1981. Boundaries of geometrical optics applicability and related problems. *Radio Sci.*, **16**(6), 975–978.

Kravtsov, Yu.A. and Orlov, Yu.I., 1983. Caustics, Catastrophes and Wave Fields. *Sov.Phys. - Uspekhi*, **26**(12), 1038–1055.

Kravtsov, Yu.A. and Orlov, Yu.I., 1990. *Geometrical optics of inhomogeneous media*. Berlin, Heidelberg: Springer Verlag.

Kravtsov, Yu.A. and Orlov, Yu.I., 1993. *Caustics, Catastrophes and Wave Fields*. Berlin, Heidelberg: Springer Verlag.

Kucherenko, V.V., 1974. Asymptotic solutions of the system $A(x_1, i\partial\hbar/\partial x)u = 0$ at $\hbar \to 0$ in the case of characteristics of variable multiplicity. *Izvestiya Akad. Nauk SSSR, Ser. Matematika*, **38**, 625–662.

Landau, L.D., 1932. On the theory of energy transfer at collisions. *ZS. Sowjet Union*, **2**(1), 46–51.

Landau, L.D. and Lifshitz, E.M., 1960. *Electrodynamics of Continuous Media*. Oxford: Pergamon.

Landau, L.D. and Lifshitz, E.M. 1970. *Theory of Elasticity*. Oxford: Pergamon.

Landau, L.D. and Lifshitz, E.M., 1977. *Quantum Mechanics: Non-Relativistic Theory*. Oxford: Pergamon.

Landau, L.D. and Lifshitz, E.M., 1987. *The Classical Theory of Field*. Oxford: Pergamon.

Law, C.T. and Kaplan, A.E., 1989. Dispersion related multimode instabilities and self-sustained oscillations in nonlinear counterpropagating waves. *Opt. Lett.*, **14**, 734–736.

Law, C.T. and Kaplan, A.E., 1991a. Instabilities and amplification of counterpropagating waves in a Kerr nonlinear medium. *J. Opt. Soc. Am.*, **8**(1), 58–67.

Law, C.T. and Kaplan, A.E., 1991b. Dispersion related amplification in a nonlinear fiber pumped by counterpropagating waves. *Opt. Lett.*, **16**, 461–463.

Lax, P.D., 1957. Asymptotic solutions of oscillatory initial value problems. *Duke Math. J.*, **24**, 627–646.

Levin, M.L. and Rytov, S.M., 1956. On the transition to geometrical approximation in the elasticity theory. *Akustich. Zhurn.*, **2**(2), 173–176 [in Russian].

Liberman, V.S. and Zel'dovich, B.Ya., 1992. Spin-orbit interaction of the photon in an inhomogeneous medium. *Phys. Rev. A.*, **46**(8), 5199–5207.

Love, J.P., Sammut, R.A. and Snyder, A.W., 1979. Birefringence in elliptically deformed optical fibres. *Electron. Lett.*, **15**(20), 615–616.

Majorana, E., 1932. Atomi orientati in campo magnetico variable. *Nuovo Cimento*, **9**(2), 43–50.

Maker, P.D., Terhune, R.W. and Savage, C.M., 1964. Intensity- dependent changes in the refractive index of liquids. *Phys. Rev. Lett.*, **12**(18), pp.507–509.

Maslov, V.P., 1972. *Théorie des Perturbations et Methode Asymptotique*. Paris: Dunod.

Maslov, V.P. and Fedoryuk, M.V., 1981. *Semiclassical Approximation in Quantum Mechnics*. Hingham: Reidel.

Melrose, D.W., 1974. Mode coupling in the solar corona: coupling near the plasma level. *Austral. J. Phys.*, **23**(1), 31–42.

Melrose, D.B. and McPhedran, R.C., 1991. *Electromagnetic Processes in Dispersive Media*. Cambridge: Cambridge University Press.

Miyano, K. and Ketterson, J.B., 1979. Sound propagation in liquid crystals. In *Physical Acoustics*, Eds. Mason, W.P. and Thurston, R.N., 93–178. New York: Academic Press.

Motz, L. and Rose, M.E., 1936. On space quantisation in time varing megnetic field. *Phys. Rev.*, **50**(3), 348–355.

Murnaghan, F.D., 1959. *Finite Deformation of an Elastic Solids*. New York: Academic Press.

Naida, O.N., 1970. On the solution of the equations of 'quasi-isotropic' approximation of geometrical optics. *Izv. VUZ: Radiofizika*, **13**(5), 751–765. [Engl. transl. in *Radiophys. and Quantum Electron.*, 1970, **13**(5)]

Naida, O.N., 1971. Corrections to the polarization of normal modes. *Izv. VUZ: Radiofizika*, **14**(12), 1843–1856. [Engl. transl. in *Radiophys. and Quantum Electron.*, 1971, **14**(12)].

Naida, O.N., 1972. On the problem of limiting polarization. *Izv. VUZ: Radiofizika*, **15**(5), 751–765. [Engl. transl. in *Radiophys. and Quantum Electron.*, 1972, **15**(5)]

Naida, O.N., 1974a. 'Sewing' of normal modes with the solutions of 'quasi-isotropic' approximation. *Izv. VUZ: Radiofizika*, **17**(6), 896–901. [Engl. transl. in *Radiophys. and Quantum Electron.*, 1974, **17**(6)]

Naida, O.N., 1974b. Method of quasi-isotropic asymptotic for electromagnetic waves in moving media. *Izv. VUZ: Radiofizika*, **17**(12), 1833–1850. [Engl. transl. in *Radiophys. and Quantum Electron.*, 1974, **17**(12)]

Naida, O.N., 1977a. Uniform geometrical optics approximation for linear systems along the rays of varying multiplicity. *Izv. VUZ: Radiofizika*, **20**(3), 383–398 [Engl. transl. in *Radiophys. and Quantum Electron.*, 1977; **20**(3)]

Naida, O.N., 1977b. Ray approximation in acoustics of inhomogeneous anisotropic media. *Doklady AN SSSR*, **236**(4), 842–845. [Engl. transl. in *Sov. Phys. - Doklady*, 1977]

Naida, O.N., 1978a. Geometrical acoustics of 3D inhomogeneous anisotropic media. *Akustich Zhurn.*, **24**(5), 731–739. [Engl. transl. in *Sov. Phys. - Acoust.*, 1978, **24**(5)]

Naida, O.N., 1978b. Geometrical optics of 3D inhomogeneous anisotropic media. *Radiotekhnika i Electronika*, **23**(12), 2489–2498. [Engl. transl. in *Sov. J. Comm. Technol. Electron.*, 1978, **23**]

Naida, O.N., 1979. 'Tangent' conical refraction in 3D inhomogeneous weakly anisotropic media. *Zhurn. Eksper. Teor. Fiziki*, **77**(2), 471–482. [Engl. transl. in *Sov. Phys. - JETP*, 1979]

Naida, O.N. and Prudkovskii, A.G., 1977. WKB method for equation $(-i\hbar\partial/\partial t + A(x, t, -i\hbar\partial/\partial x)U = 0$ in the case of characteristics of varying multiplicity. *Differentsialnye uravnenia*, **13**(9), 1678–1691. [Engl. transl. in *Differential equations*, 1977, **13**(9)]

Naida, O.N. and Prudkovskii, A.G., 1978. Quasi-classical description of neutron polarization in inhomogeneous magnetic field. *Yadernaya Fizika*, **28**(6), 1560–1568. [Engl. transl. in *Nuclear Physics*, 1978, **28**]

Oldano, C., 1987. Electromagnetic wave propagation in anisotropic stratified media. *Phys. Rev. A.*, **40**(10), 6014–6022.

Olver, F.W.J., 1954. The asymptotic expansion of Bessel functions of large order. *Phil. Trans. Roy. Soc.*, (Lond.) **A247**, N930, 328–368.

Olver, F.W.J., 1958. Uniform asymptotic expansions of solutions of linear second-order differential equations for large values of a parameter. *Phil. Trans. Roy. Soc.*, (Lond.) **A250**, N984, 479–518.

Pauli, W., 1932. Diracs Wellengleichung des Electrons und geometrishe Optik. *Helv. Phys. Acta.*, **5**(3), 179–199.

Payne, D.N., Barlow, A.J. and Ramskov-Hansen, J.J., 1982. Development of low- and high-birefringence optical fibers. *IEEE J. Quantum Electron.*, **18**(4), 477–488.

Rabi, I.I., 1935. On the process of space quantization. *Phys. Rev.*, **49**(3), 324–328.

Rashleigh, S.C., 1983. Origins and control of polarization effects in single-mode fibers. *J. Lightwave Technol.*, **1**(2), 312–331.

Rashleigh, S.C. and Stolen, R.H., 1983. Preservation of polarization in single-mode fibers. *Lazer focus*, **19**(5), p.155–161.

Ross, J.N., 1984. The rotation of the polarization in low birefringence monomode optical fibres due to geometric effects. *Opt. and Quantum Electr.*, **16**(5), 455–461.

Rubinow, S.I. and Keller, J.B., 1963. Asimptotic solution of the Dirac equation. *Phys. Rev.*, **131**(6), 2789–2796.

Rytov, S.M., 1938. On the transition from wave to geometrical optics. *Doklady AN SSSR*, **18**(2), 263–266 (in Russian).

Schiller, R., 1962a. Quasi-classical theory of a relativistic spinning electron. *Phys. Rev.*, **125**(3), 1116–1123.

Schiller, R., 1962b. Quasi-classical theory of a relativistic spinning electron. *Phys. Rev.*, **128**(3), 1402–1412.

Smith, A.M., 1980. Birefringence induced by bends and twists in single-mode optical fiber. *Appl. Opt.*, **19**(15), 2606–2611.

Snyder, A.W. and Love, J.P., 1983. *Optical Waveguide Theory*. London, New York: Chapman and Hall.

Sobelman, I.I., 1992. *Atomic spectra and radiative transitions*. Berlin, Heidelberg: Springer Verlag.

Sobelman, I.I., Vainstein, L.A. and Yukov, E.A., 1995. *Exitation of atoms and broadening spectral lines*. Berlin, Heidelberg: Springer Verlag.

Stern, O., 1921. Ein Weg Zur experimentallen Prufung der Richtungsquantelung im Magnetfeld. *ZS. f. Phys.*, **7**(2), 249–258.

Stepanov, N.S. and Gavrilenko, V.G., 1971. On the electromagnetic wave theory for weakly nonstationary plasma. *Doklady AN SSSR*, **201**, 507–513. [English transl. in *Sov. Phys. - Doklady*, 1971]

Suvorov, E.V., 1972. Electromagnetic wave propagation in plasma with the magnetic field shear. *Izv. VUZ: Radiofizika*, **15**(9), 1320–1324. [Engl. transl. in *Radiophys. and Quantum Electron.*, 1972, **15**]

Sverdlov, Yu.L., Miroshnikova, T.V. and Sergeeva, N.G., 1989. The small-scale fluctuations height distribution measuring method. In *Rasprostranenie radiovoln v vozmushchenoi ionosphere* (Radiowave propagation in a perturbed ionosphere), 30–35. Apapity: Kolsk. Filial AN SSSR, [in Russian].

Titcherige, J.E., 1971. Magnetoionic mode coupling of HF radiowaves. *J. Atm. Terr. Phys.*, **33**(10), 1533–1541.

Thom, R., 1972. *Structural stability and Morphogenesis*. New York: Benjamin.

Tjaden, D.L.A., 1978. Birefringence in single-mode optical fibres due to core ellipticity. *Phillips J. Res.*, **33**(5/6), 254–263.

Tokar, V.G., Rubinshtein, L.I. and Nikitin, M.A., 1987. The study of short radiowave depolarization in the ionosphere in quasi-isotropic approximation. *Izv. VUZ: Radiofiz.*, **30**(1), 36–41. [Engl. transl. in *Radiophys. and Quantum Electron.*, 1987, **30**(1)]

Tomita, A. and Chiao, R.Y., 1986. Observation of Berry's topological phase by use of an optical fiber. *Phys. Rev. Lett.*, **57**(8), 937–940.

Toupin, R.A. and Bernstein, B., 1961. Sound waves in deformed perfectly classic materials: acoustoelastic effect. *J. Acoust. Soc. Amer.*, **33**(2), 216–225.

Tromp, J., 1993. Support for anisotropy of the Earth's inner core from free oscillations. *Nature*, **366**(6456), 678–681.

Unger, H.-G., 1977. *Planar Optical Waveguides and Fibers*. Oxford: Clarendon Press.

Varhnam, M.P., Payn, D.N., Barlow, A.J. and Birch, R.D., 1983. Analytical solution for the birefringence produced by thermal stress in polarization maintaing optical fibers. *J. Lightwave Technol.*, **1**(2), 332–339.

Vacek, R., 1995. Chirality in nonlinear media. PIERS'95, 24–28 July, Seatle, WA.

Wazow, V., 1965. *Asymptotic expansions for ordinary differential equations*. New York: Interscience.

Whittaker, E.T. and Watson, G.N., 1965. *A Cource of Modern Analysis.* 4th ed., reprinted. Cambridge: University Press.

Zaitsev, Yu.A., Kravtsov, Yu.A. and Yashin, Yu.Ya., 1968. On the transition to geometrical optics approximation in electrodynamics of inhomogeneous anisotropic media. *Izv. VUZ: Radiofizika,* **11**(12), 1802–1811. [Engl. transl. in *Radiophys. and Quantum Electron.,* 1968, **11**(12)]

Zener, C., 1932. Non-adiabatic crossing of energy levels. *Phil. Trans. Roy. Soc., (Lond.),* **A137**, 696–702.

Zheleznyakov, V.V., 1970. *Radioemission of the Sun and Planets.* Oxford: Pergamon.

Zheleznyakov, V.V. and Zlotnik, E.Ya., 1963. On the polarization of radiowaves propagating through the transvers magnetic field region in solar corona. Astronomich. Zhurn., **40**(4), 633–642 [Engl. transl. in *Astronom. J.,* 1963, **40**(4)]

Zheleznyakov, V.V. and Zlotnik, E.Ya., 1977. Radio wave propagation in the inhomogeneous magnetic field of the solar corona. *Izv. VUZ: Radiofizika,* **20**(9), 144–1461. [Engl. transl. in *Radiophys. and Quantum Electron.,* 1977, **20**(9)]

Zheleznyakov, V.V., Kocharovskii, V.V. and Kocharovskii, Vl.V., 1980. linear wave coupling in the optics of liquid crystals. *Zhurn. Eksp. Teor. Fiziki,* **79**(5), 1735–1758. [Engl. transl. in *Sov.Phys. - JETP,* 1980]

Zheleznyakov, V.V., Kocharovskii, V.V. and Kocharovskii, Vl.V., 1983. Linear interaction of electromagnetic waves in inhomogeneous weakly anisotropic media. *Sov. Phys. - Uspekhi,* **26**(10), 877–902.

Zheludev, N.N., 1989. Polarization instability and multistability in nonlinear optics. *Uspekhi Fizich. Nauk,* **157**(4), 683–717. [Engl. transl. in *Sov. Phys. - Uspekhi,* 1989, **32**].

Zhang, B. and Wang, K., 1996. Theoretical study of perturbation method for acoustical multipole logging in anisotropic formation. *J. Acoust. Soc. Amer.,* **99**(5), 2674–2685.

Index

Printed and bound by CPI Group (UK) Ltd, Croydon, CR0 4YY

24/10/2024

01778279-0002